SpringerBriefs in Business

For further volumes:
http://www.springer.com/series/8860

Diana Derval · Johan Bremer

Hormones, Talent, and Career

Unlock Your Hormonal Quotient®

Springer

Diana Derval
DervalResearch
Chicago IL
USA

Johan Bremer
Bremer Studios
Amsterdam
The Netherlands

ISSN 2191-5482
ISBN 978-3-642-25712-4
DOI 10.1007/978-3-642-25713-1
Springer Heidelberg New York Dordrecht London

e-ISSN 2191-5490
e-ISBN 978-3-642-25713-1

Library of Congress Control Number: 2012935099

© The Author(s) 2012

This work is subject to copyright. All rights are reserved by the Publisher, whether the whole or part of the material is concerned, specifically the rights of translation, reprinting, reuse of illustrations, recitation, broadcasting, reproduction on microfilms or in any other physical way, and transmission or information storage and retrieval, electronic adaptation, computer software, or by similar or dissimilar methodology now known or hereafter developed. Exempted from this legal reservation are brief excerpts in connection with reviews or scholarly analysis or material supplied specifically for the purpose of being entered and executed on a computer system, for exclusive use by the purchaser of the work. Duplication of this publication or parts thereof is permitted only under the provisions of the Copyright Law of the Publisher's location, in its current version, and permission for use must always be obtained from Springer. Permissions for use may be obtained through RightsLink at the Copyright Clearance Center. Violations are liable to prosecution under the respective Copyright Law.

The use of general descriptive names, registered names, trademarks, service marks, etc. in this publication does not imply, even in the absence of a specific statement, that such names are exempt from the relevant protective laws and regulations and therefore free for general use.

While the advice and information in this book are believed to be true and accurate at the date of publication, neither the authors nor the editors nor the publisher can accept any legal responsibility for any errors or omissions that may be made. The publisher makes no warranty, express or implied, with respect to the material contained herein.

Printed on acid-free paper

Springer is part of Springer Science+Business Media (www.springer.com)

To us, 28.04.2012

Foreword

When I read Diana's new book, I discovered it is not only novelists who can make you live the story and feel that you are part of it: from the first line to the last word I could not stop reading. Her search for the truth of hormones and people reminds me of the journey of the boy Santiago following his dreams in his search for the secrets of wisdom in Paulo Coelho's grand novel *The Alchemist*.

Both are deeply involving: while Santiago's search is mythical, Diana's and Johan's is scientific. They brilliantly answer the questions: Are we born as leaders or entrepreneurs? Are our physical traits, sensory perception, and personality determined in the womb? What is the effect of our hormones on our behavior, career choices, success, and even love life? What path shall we take in life and what makes us who we are?

A fascinating study of the effect of hormones on the leadership style of recognized business leaders and great movie directors. Once again Diana impresses me with her deep analysis and broad knowledge. In this book she helps you to unlock the closed doors of your Hormonal Quotient®, and to read—as what Paulo Coelho quoted in my native Arabic language *Maktub* meaning—*what's written in it*.

Shawqi Ghanim
General Manager, Grand Optics

Acknowledgments

Writing this book was a great adventure and we would like to thank all the people who helped us on the way.

A warm thank you to the thousands of people, professionals, DervalResearch teams and clients, and students from ESSEC Executive Education, Robert Kennedy College, and University Leonard de Vinci, who participated with our research.

We are particularly grateful to Shawqi Ghanim (Grand Optics) for sharing his experience and accepting to write the wonderful foreword of the book. We thank Robert-Jan Woltering (Sofitel Legend Amsterdam The Grand), Bengt Järrehult "Dr. Beng" and Donna Martin (SCA), Magda Carvalho (Patent Attorney), Ariane Latreille (L'Oréal CCB), Nick Goldie (Philips), Wai Wong (Sephora), Virgine Bellière-Baca (Solvay), Hilary Ellis (EuropeanPWN), Stefanie Schiller and Matthias Krämer (Siemens Healthcare), Raven Hanna (Made with Molecules) for sharing their experience and supporting our research.

This book would look like nothing without the editing performed by David Gardner, the drawings by Vlad Kolarov, the face analysis powered by Bremer Studios, the preliminary research conducted by Marja Salaspuro, the portrait photographs taken by Annabelle, and the book trailer shot by Jochem and Carlos.

The research previously done by John T. Manning on prenatal hormones, Geert Hofstede on cultural differences, Marvin Zuckerman on sensation seeking, Bronwen Martin on metabolism, and Elaine N. Marieb on physiology was a fantastic foundation for our work. Thank you so much for sharing your expertise!

Special thanks to Lara for her help with the Better Immune System Foundation and to TomTom teams. Bedankt to our family and friends, especially Sandrine & Nick, Jeff Phrase, Nicolas, RJ & Yunnao, Edwin, Anna & Mireille, Lizet & Robert.

Vielen Dank to Barbara Fess for her trust and guidance.

And many thanks to you dear reader!

Diana & Johan

Reviews

"Throughout this book Diana clearly explains what makes us different as individuals. Understanding my HQ (Hormonal Quotient) helped me to have a better understanding of who I am, how I interact with others, and why I am great or not in various areas. I recommend this book to anyone who wants to improve the way he/she apprehends things at work and in life in general."

– Vincent Harper, owner at Webkumo, Sales Order Flow Coordinator at Levi's

"Packed with funny yet mind-blowing anecdotes, this book reconciles our biological and our intellectual selves. Translating scientific facts into useful career management tips like how to maximize your strengths or optimize your potential, this book will surprise you and put you to the test—by the way, I really enjoyed the Monkey/Bear/Banana test! This book will make you frown sometimes—What? Really? Is that so?—and will always enlighten you with its findings. Well documented but never boring, it will be a truly useful read, whether you want to challenge your current career path, understand why you don't seem to enjoy the same hobbies as your spouse, or just get a fun-packed trip through the amazing world of the human body and mind."

– Sandrine Goldie, Finance Manager at British American Tobacco

"Stop looking for a coach and invest in the newly released 'Hormones, Talent, and Career' book written by the talented Diana Derval. Besides helping you in understanding which job will fit best your Hormonal Quotient profile, it also takes into consideration critical dimension such as recognition, competition, and interaction. It will also help you identify the country and type of organization that fit you best. This book is the guide for your next career move!"

– Brigitte Baladié, Business Development Manager at Intralox

"Once again Diana impresses me with her deep analysis and broad knowledge! In this book she helps you to unlock the closed doors of your Hormonal Quotient.

A fascinating study of the effect of hormones on the leadership style of recognized business leaders."

– Shawqi Ghanim, General Manager at Grand Optics

"This book is to talent management what iPhone is to communication: innovative, simple, and convenient. A must-have! Professor Diana Derval and Johan Bremer developed with passion and patience this unexpected scientific vision. Their innovative approach to career management based on hormonal factors will not leave managers, leaders, coaches, and HR managers indifferent. This useful and playful book can be used in both a management and a personal development context. I personally enjoyed reading the ebook version."

– Cecilia Laustriat, Head of International Digital Marketing at BNP Paribas Real Estate, President of the Essec Executive MBA Alumni Association

"Diana Derval and Johan Bremer have written the most innovative book about building and creating teams! Their unconventional research showed that hormones influence team members and thus results. Although it is a biological subject, the writing style is easy with lots of personal examples. The book has the potential to inspire managers to reach greater heights."

– Lizet Elshof, Unit Manager at Lloyd's Register

"Beware. This book will change your life! You might quit your job or divorce after reading it. But you will understand why you are the way you are."

– Natalie Ardet, owner at ArdetConsult, Project Manager at Vodafone

"I love this book! All relevant information on personal development gathered in a hardcover. I couldn't stop reading. The chapter about how we pick people at work and in our private life was a real eye opener to me."

– Diana Eilert, Manager Product Development at SCA Hygiene Products

"Diana Derval and Johan Bremer are offering an innovative way of looking into who we are, and most importantly what we can do with what we are, from a professional and personal perspective. The combination of the rich content and thoroughly documented examples presented with lively and fun writing will make it difficult for you to stop reading this book once you get started, making you eager to know more after every page. And when you are done, even more eager to go out there and share and test in the real world! An interesting point is that by reading this book, you also understand who the authors are and what actually drives their own creativity and ability to think "out of the box"."

– Nicolas Blaisot-Balette, Senior Manager Innovation, Novedia Group

"There are well-known examples, like The Game of Life, that demonstrate how all kinds of complexity arise from a couple of rules and simple prior conditions of the "living cells" - the building blocks of the Game of Life Universe. The new book "Hormones, Talent and Career" written by Diana Derval and Johan Bremer, appears to state something similar, albeit in a more touchy domain: that many aspects of complexity of human behavior might be influenced by hormonal "building blocks", a chemical "signature" that is unique for every person. A number of studies, demonstrating correlation between the presence or absence of a particular hormone and various behavioral patterns, might urge a reader to check their own hormonal "signature" - perhaps, getting some fresh insights into their own life story. An interesting journey to the crossroads between the human psyche and the biological self."

– Anna Nachesa, Software Engineer at eBay

Contents

1	**We are Our Hormones**	1
	Pushing Buttons	1
	Aren't You Overreacting?	1
	Binding Potential	2
	Hormonal Disruptors	3
	What is Written in Our DNA?	3
	The Protein Compiler	4
	A Unique Source Code	4
	And the Winner Is	5
	The Double Effect of Hormones	5
	Everything is Played in the Womb!	6
	Gender Polymorphisms	7
	The Hormonal Quotient® (HQ)	8
	References	9
2	**Unlock your Hormonal Quotient® (HQ)**	11
	Seven Things That Really Matter in a Job	11
	Movement: Sensation or Routine?	11
	Interaction: Object or Person?	13
	Competition: Fight or Flight?	15
	Recognition: Status or Feedback?	17
	Structure: Analogical or Sequential?	18
	Collaboration: Input or Co-Creation?	20
	Find Your Own Path	21
	Biomarkers and Hormonal Quotient® (HQ)	22
	What You See Is What You Get	22
	How Bankable Are You?	23
	Our Future Is in Our Hands	26
	Measure Your Hormonal Quotient® (HQ)	27
	References	30

3	**Find the Right Career Path**	35
	Physiology and Job Fit	35
	Hormonal Quotient® (HQ) and Occupation in Women	35
	Vision and Career	37
	Friendly Work Environment	37
	Hormonal Quotient® (HQ) and Ideal Job	40
	Very-Testosterone HQ: BU Manager versus Accountant	40
	Testosterone HQ: Product Manager versus Doctor	42
	Balanced HQ: Operations Director versus Surgeon	45
	Estrogen HQ: Chief Information Officer versus Politician	48
	References	51
4	**Build a Winning Team**	53
	Hormones, Entrepreneurship, and Leadership	53
	Entrepreneur or Top Manager?	53
	Same Job, Different Styles: The Movie Directors' Case	57
	Ideal Country and Type of Organization	59
	New Strategies to Recruit, Motivate, and Lead Teams	64
	The Clone Wars: The Sales Case	64
	Recruiting and Motivating the Right People	67
	Hormonal Quotient® and Leadership Style	69
	References	69
5	**Find the Perfect Balance**	71
	Hormonal Quotient® (HQ) and Lifestyle	71
	Hormones, Thyroid, and Ayurveda	71
	Discover Your Hidden Talents	74
	Find the Right Lifestyle	75
	Hormonal Quotient® (HQ), Friends, and Love	75
	Perfect Buddies	75
	Hormones and Family Life	76
	Find Your Missing Half	77
	References	79
	Conclusion	83

About the Authors

You can contact Diana Derval at diana@derval-research.com or via the website www.derval-research.com and follow her on twitter@profdianaderval

Diana Derval is President and Research Director of DervalResearch, global market research firm specializing in human perception and behavior, and Chair of the Board of Directors of the Better Immune System Foundation. Inventor of the Hormonal Quotient® (HQ), member of the Society for Behavioral Neuroendocrinology, and author of the books "Wait Marketing" and "The Right Sensory Mix", Diana turns fascinating neuroscientific breakthroughs into powerful business frameworks to identify, understand, and predict human traits, motivations, and behavior. She has accelerated the development of Fortune 500 firms including TomTom, Michelin, Sofitel, Philips, Sara Lee, and Sephora. Diana Derval teaches Innovation and Sensory Science at ESSEC Paris-Singapore Business School, and at the MBA of the University Leonard de Vinci in Paris. Over 10,000 professionals have enjoyed her inspirational lectures and workshops from Chicago to Shanghai.

DervalResearch

You can contact Johan Bremer at johan@bremerstudios.com

Johan Bremer, MSc in computer science, is the owner and technology director of Bremer Studios, a firm with expertise in 3D rendering and scientific visualization. Johan and his team have implemented in the past decade must-have applications for leading organizations, including lane information on TomTom GPS navigation devices, a 3D visualization of the Anne Frank Museum tree based on a laser-scanning dataset, and neuromarketing demonstrations for DervalResearch and Philips. Combining strong mathematical skills with an ability to quickly grasp needs of demanding customers, Johan Bremer also expanded the visual program Crystal Ball, an option-risk application for traders.

By the Same Authors: The Right Sensory Mix

"*Diana Derval has written the best book that I have seen on the critical role of the five senses in determining our brand preferences. Her writing is lively, full of relevant case studies, and rich in insights. No marketing department or new product department must proceed without first reading this book.*"
- Philip Kotler, S.C. Johnson & Son Professor of International Marketing, Kellogg School of Management

Why do some people drink black coffee and others stick to tea? Why do some people prefer competitors' products? Why do we sell less in this country?

Many companies fail to acknowledge and analyze disparities observed among customers and simply put them down to culture or emotion. New neuroendocrinological research proves that consumers are rational: They just have a different biological perception of the same stimulus! Their preferences, behavior, and decisions are strongly influenced by the hundreds of millions of sensors monitoring their body and brain. People with more taste buds are for example sensitive to bitterness and are more likely to drink their coffee with sugar or milk, or to drink tea.

After reading the book, managers will be able to:

- Understand and predict consumers' behavior and preferences
- Design the right sensory mix (color, shape, taste, smell, texture, and sound) for each product
- Fine-tune their positioning and product range for every local market
- Systematically increase their innovation hit rate

Diana Derval and Johan Bremer support the Better Immune System Foundation.

> Many immune system disorders like eczema, asthma, sinusitis, hyperacusis, diabetes, autism, and more are in fact related to our hormonal balance. The mission of the Foundation is to conduct research, information, and prevention programs for a Better Immune System.

As an individual or a corporation, you can select the programs you wish to support at www.betterimmunesystem.org

We thank you for your contribution.

Introduction

In the past years, five hundred scientific articles have been published analyzing the influence of hormones on brain and behavior. Media are now regularly featuring breakthroughs in neuroendocrinology, such as the link between prenatal hormones and financial performance, competitiveness, or risk management.

The role of hormones indeed goes far beyond male/female differences, swinging moods, or acne rashes: hormones have their say regarding the way we look, the way we interact, and even the way we live. They even override gender and culture.

Recent papers have demonstrated an impact of prenatal hormones on our body mass index, shape of our face, length of our fingers, ability to articulate, to use a variety of words ("stuff" and "thing" do not count), and to write in a legible way.

Our emotional reactions, level of anxiety, possible hyperactivity, social skills, co-operative behavior, and sensation seeking traits are also related to hormones. Whether we would be great athletes, firefighters, professional rugby or football players, as well as our level of performance can be predicted. Our ability to play music or to aim for a target, our skiing or rowing times are all set from birth.

The same applies to our family life. The number of sex partners we have, the length of the relationship, and even the size of our future family is played in advance.

We also know from which disease we are more likely to suffer: osteoarthritis, prostate cancer, myocardial infarction, autism spectrum disorder, anorexia nervosa, and amiotrophic lateral sclerosis (ALS) are all linked to the influence of hormones in the womb. Appetite, eating disorders, and dependence on alcohol follow the same scheme.

All the rest we do is 100% based on our free will.

In our previous book "The Right Sensory Mix" we revealed the impact of hormones on sensory perception (smell, touch, taste, vision, hearing) and products preferences. We also unveiled some personality traits related to innovation and leadership. It is now time to reveal the role of hormones on our social skills, talents, and career choices.

Unlock Your Hormonal Quotient®

Based on these publications and on the measurements we performed on 3,500 people in over 25 countries, we identified eight Hormonal Quotient® (HQ) profiles, and the very specific career paths, hobbies, and lifestyle that best suit each profile. An in-depth analysis of the data collected and actionable recommendations will help managers benchmark their Hormonal Quotient® (HQ) with peers, design an ideal career plan for themselves and for their teams, and unlock their full potential.

This book will help managers leverage this groundbreaking knowledge to boost their career and also recruit, motivate, and lead teams to success. The strength of the presented approach is to use biomarkers in order to accurately identify our talent, and also compatible jobs, hobbies, and lifestyle.

We are our hormones. What about you?

Chapter 1—**We Are Our Hormones**

Explains the influence of hormones on our personality traits, physical aptitudes, skills, and preferences. The link between genes, hormones, and environment is studied. The concept of gender polymorphism is applied to humans, introducing the Hormonal Quotient® (HQ).

Chapter 2—**Unlock Your Hormonal Quotient® (HQ)**

Presents the seven things that really matter in a job. Reviews main biomarkers like shape of the face, and digit ratio, and what they reveal about the personality traits of emblematic business leaders such as Steve Jobs, Mark Zuckerberg, Barack Obama, and Donald Trump. Explains how to measure the Hormonal Quotient® (HQ). Presents the typical traits for each of the eight Hormonal Quotient® (HQ) profiles.

Chapter 3—**Find the Right Career Path**

Presents the ideal jobs, and also career paths to avoid, depending on our physiology and Hormonal Quotient® (HQ). Reveals the talents and Hormonal Quotient® (HQ) of managers from leading brands, like L'Oréal, Philips, Sofitel, and Solvay and shares how they used their potential to build a successful career.

Chapter 4—**Build a Winning Team**

Presents the ideal type of organization and country depending on our Hormonal Quotient® (HQ). Recommends strategies to recruit, motivate, and lead teams based on the Hormonal Quotient® (HQ) of each member. Lists strengths and points to improve for each type of leader.

Chapter 5—**Find the Perfect Balance**

Reviews the ideal hobbies, sports, nutrition, family life, friends, and missing half for each Hormonal Quotient® (HQ) profile. Shares the Hormonal Quotient® (HQ) and lifestyle of celebrities such as Brad Pitt and Johnny Depp. Reveals the best love compatibility between Hormonal Quotient® (HQ) profiles.

Chapter 1
We are Our Hormones

> *The physiology of today is the medecine of tomorrow.*
> Ernest Starling, English physiologist (1866–1927), discovered the first hormone in 1905

Whether we like the idea or not, hormones are pretty much shaping who we are. Our physical traits, sensory perception, personality, and more, are determined in the womb. On the bright side, once we are aware of the perimeter of our abilities and skills, we can focus on the few aspects (luckily there are still a couple of them!) we can actually change.

In this chapter, we will see how hormones, genes, and environment interact. What is written in the deoxyribonucleic acid (DNA)? What happens in the womb? Does the environment have any influence? All the answers are here.

Pushing Buttons

Hormones are the guys pushing buttons in our body: these molecular triggers turn on and off functions pre-programmed in our cells. For instance, cortisol, a stress hormone, will in case of perceived danger redirect the blood flow towards critical organs like the heart to optimize our reaction time.

Hormones are messengers. They activate or inhibit target cells (Fig. 1.1). Some hormones like insulin, oxytocin, or vasopressin, are messenger proteins. Some, like dopamine and serotonin, are called neurotransmitters as they transfer signals from a neuron to a target cell. Sex hormones like estrogen, testosterone, and progesterone are key in the sexual differentiation process.

Aren't You Overreacting?

If your entourage keeps wondering "Aren't you overreacting?", here are some facts you can share with them:

Fig. 1.1 Hormonal response

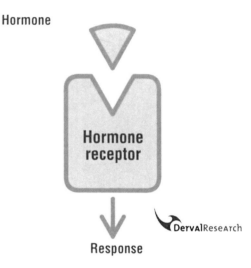

In case of stress for instance, the intensity of our reaction depends on:

(a) the level of *circulating hormones* in our blood (here the level of cortisol),
(b) the number of *receptors for that hormone* located in the cells (cortisol can bind with protein receptors),
(c) the *affinity* between the hormone and the receptor, how well they bond together, also called *binding potential*

This explains why in a similar situation people might react very differently.

Binding Potential

The binding potential is defined by the density of the receptors, their affinity, and the free fraction (of not bound hormones) (Varrone 2011).

The technique used to measure the binding potential for a hormone or substance is a dynamic positron emission tomography (PET) scan. The idea is to select a substance that binds to a receptor. Then to follow the distribution of this tracer in a specific location of the body and brain over time. And then calculate and visualize how much of the tracer bound to the receptors and which fraction is still free.

With this shot (Fig. 1.2) taken by the Siemens Healthcare's Biograph molecular magnetic resonance (mMR) device, you can see at the same time mMR sequences and dynamic PET data of the studied location—a world première!

The Biograph is an imposing device that looks a bit like a professional pizza oven (with us being the actual pizza!). We would use this accurate device in a clinical environment only. We will see in Chap. 2 alternative and non-invasive techniques to evaluate our response to hormones in our living room.

Fig. 1.2 Head scan (Siemens Biograph mMR)

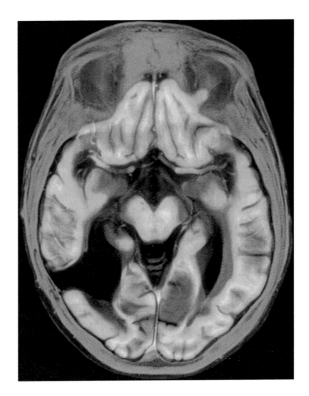

Hormonal Disruptors

Hormones receptors can be messed up by internal and external disruptors.

An example of an internal disruptor could be estrogen: the molecule is similar to aldosterone, the hormone that regulates potassium K+ and sodium NA+ intake. Confusion happens for instance when estrogen levels increase during menstruation or pregnancy—women in that situation tend to retain too much fluid, causing edema (Marieb and Hoehn 2007, Unit 4 Chapter 26).

External disruptors can be any kind of chemical (such as fragrances, food, pollutants, and drugs). As you can see in the pictures below, the receptors bind with odorants and food in the exact same way they bind to hormones (Fig. 1.3), in a Tetris way.

What defines the number of receptors we have? Is it written in our DNA? Shaped by hormones? All the answers now.

What is Written in Our DNA?

Let's first understand how genes and hormones work together to create "moi". What is written in the DNA? What is hereditary? Are we unique?

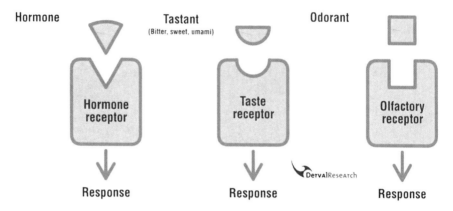

Fig. 1.3 Hormone, tastant, odorant response

The Protein Compiler

The human body is a protein compiler. These complex molecules—mostly made of carbon, oxygen, hydrogen, and nitrogen—play a critical role in the construction and function of body and brain (National Institutes of Health 2012a).

Each protein has a role. Collagen, for instance, makes the skin elastic, while keratin makes it waterproof. Actin and myosin take care of muscle contractions. Insuline regulates the level of sugar. Immunoglobulins, our internal cleaning Mo for those who remember Wall-e, identify "foreign contaminants" like viruses or allergens (Wall-e movie). Even growth hormone is a protein (Marieb and Hoehn 2007, Unit 1 Chapter 2).

Good to know: proteins are sensitive folks—they are unable to perform under extreme temperatures (now I know why my skin is dryer in winter!) or unstable pH. We will come back to this in Chap. 5 on nutrition and health.

A Unique Source Code

Our DNA (deoxyribonucleic acid)—located in the nucleus of each cell—is the source code where everything is written. When the cells divide, they copy and paste this code and pass it equally to the daughter cells so that they will be able to transcribe and compile it into proteins. Individuals of the same species share 99.9% of the same DNA. Which means that variations in only 0.1% of the DNA are enough to make us unique. The chances to meet our clone is one in a trillion. We will see in Chap. 4 that it happens more often than we would think.

DNA—organized into chromosomes—is coded in C, and also in A, G, and T. The particular combination of C (for cytosine), A (for adenine), G (for guanine), and T (for thymine) determines whether we will be a human, a lizard, or a cactus.

A single variation of C, A, G, or T, in any location (or locus—always more fun in Latin) can translate into very specific traits or skills (Marieb Unit 5 Chapter 29). This single variation is called a SNP (Single Nucleotide Polymorphism—pronounce "snip"). Each of these SNPs is a biomarker that can help predict our reactions (National Institutes of Health 2012b).

> **Are You Dry or Wet?**
> Humans can have a wet or a dry type of earwax. We can find the dry type in Asia for instance. A SNP is actually responsible for defining our earwax. In the locus 538G – > A (rs17822931), gene ABCC11, a AA SNP leads to dry earwax, whereas a GA or GG characterizes the wet type of earwax. Useful to know in order to pick the right cleaning strategy (Yoshiura et al. 2006).

And the Winner Is

Each chromosome contains one gene (or allele) either from the father or from the mother. We might receive contradictory instructions like "blue eyes" or "brown eyes", so who wins? The dominant allele of course. Recessive allele must be paired with an identical recessive allele in order to bring their voice to the chapter. In that example, blue eyes are recessive. It's also the case for sensitivity to bitterness—measured with PTC strips, an organic compound called phenylthiocarbamide, that tastes bitter for *medium-tasters* and *super-tasters*—people equipped with many taste buds—and just like paper for *non-tasters*—people with fewer taste buds (Derval 2010). If someone is a *non-taster* and is less sensitive to bitterness, both parents had to pass a *non-taster* gene. We will see in Chap. 4 the fascinating consequences on team recruitment. Thin lips are also recessive. An example of a dominant feature is dimples in the cheeks (Marieb and Hoehn 2007, Unit 5 Chapter 29). Only one parent needs to transmit it in the genotype (genes) and it is directly expressed in what is called the phenotype (National Institutes of Health 2012c).

Note that if your children are taller than you, they might still be your natural descent. Some attributes like height are determined by a combination of genes so that the inherited outcome might be less directly expected.

So that would be it: our genes dictate how we will end up or start. This is without taking into account the double effect of hormones shaping our body and mind and pre-programming our future reactions.

The Double Effect of Hormones

Exposure to prenatal hormones has immediate and lasting effects on our body and brain.

Fig. 1.4 In the womb

Everything is Played in the Womb!

While in the womb (Fig. 1.4), men or women, we are all exposed to hormones like testosterone, estrogen, and cortisol but at different stages of our fetal growth and with different dosages.

These hormones like glucocorticoids clearly regulate fetal growth. There is 10 times more glucocorticoids circulating in the mother than in the fetus but it varies among individuals. Glucocorticoids determine the density of hormones receptors, and activate or inactivate other hormones. Newborns with higher basal cortisol levels are more likely to develop hypertension and diabetes at a later stage. The concentration also depends on the fetal barrier, and external stimuli like nutrition or stressful events.

Exposure to sex hormones during a specific and even narrow window of time can have lasting effects (Fowden and Forhead 2004). Prenatal hormones are responsible for the sexual differentiation of the body and of behavior. For instance, female monkeys exposed to testosterone very early in the fetal development enjoy more rough play than grooming (Cohen-Bendahan et al. 2005).

These prenatal hormones are critical as they have an *organizational effect* on our body and brain: they determine our physical and personality traits while we are still a fetus, and this is independent of our gender (Pinel 2007). We already know before the 14th week if we will have generous breasts, be bald, or competitive (Manning 2002).

The hormones also have an *activational effect* on our body and brain that conditions our reactions throughout our life. Prenatal glucocorticoids like cortisol (that are derived from progesterone) change the expression of hormone receptors present in our cells and influence the way genes will be transcribed in the future and pre-define the intensity and nature of our reactions. These indirect effects of hormones are often wrongly attributed to psychological or cultural factors.

The pairs of genes we receive can be totally washed away during intra-uterine programming, in the womb. One thing is sure: We are our hormones.

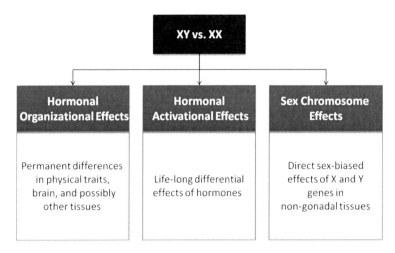

Fig. 1.5 Double effect of hormones

Gender Polymorphisms

In many animal species, including birds, fishes, and lizards, you can find more than two genders. I am not talking here about drag-chickens, but about different expressions of male and female, called gender polymorphism. These multiple genders are easily detected in some species where each group has a distinctive appearance.

For instance in the side-blotched lizard there are 3 male and 2 female genders (Roughgarden 2004):

- orange-throated males, aggressive, with high level of testosterone, and dating several females
- blue-throated males, less testosterone, less aggressive, taking care of one female at a time
- yellow-throated males, with no fixed territory
- orange-throated females, laying many eggs (5.9 per batch), distant to other females
- yellow-throated females, laying fewer but bigger eggs (5.6 per batch), more tolerant with other females

Interestingly, by injecting progesterone into a lizard the day it hatches from the egg, researchers could determine whether the lizard would be blue or yellow-throated. The gender polymorphism determines the physical traits and the behavior of the individuals, and is directly linked to prenatal hormones.

The theory is that X and Y genes influence gender expression directly via sex chromosome effects, but also indirectly via hormonal effects (Fig. 1.5). These hormonal effects can be 'organizational' or 'activational' as we saw (Arnold et al. 2009).

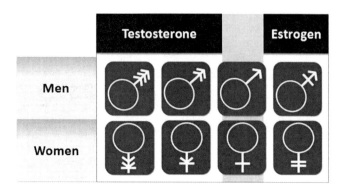

Fig. 1.6 Gender, and Hormonal Quotient® (HQ)

In the example of our tree lizards, the action of prenatal hormones 'organized' their body and brain. This is why the lizard turned into an orange-blue male after exposure to progesterone. Other behaviors are 'activated' later, like when orange males have an increase in cortisol due to the stress of being thirsty and decide to become nomadic to seek for food and beverages whereas orange-blue lizards never perceive similar stress signals even when the fridge is empty (Roughgarden 2004).

The Hormonal Quotient® (HQ)

When we started research on human behavior a couple of years ago, we felt that the traditional men/women segmentation didn't help at all to explain observed phenomena. Inspired by the research done on gender polymorphism in animals, we formulated a hypothesis that there were more than one type of men and women. We confronted our hypothesis to the measurement on 3,500 people in over 25 countries between 2007 and 2011. We were able to identify 4 gender polymorphisms for men, and 4 for women, depending on the organizational effects of prenatal testosterone and estrogen (Fig. 1.6). People of each group presented similar perception, talents, and behavior. We called the influence of hormones on our traits, behavior, and sensory perception our Hormonal Quotient® (HQ).

Women can be equally influenced by testosterone and estrogen—they are represented by the standard female symbol we are familiar with. They can also be estrogen-driven (with two stripes instead of one), or on the other side testosterone-driven (with an arrow), or very testosterone-driven (with two arrows).

Men can be equally influenced by testosterone and estrogen—they are represented by the standard male symbol we are familiar with. They can also be testosterone-driven (with two arrows instead of one), or very testosterone-driven (with three arrows), or on the other side estrogen-driven (with a stripe).

With strong variations between countries, we observed that 50% of the men and women are equally influenced by testosterone and estrogen, that 20% of the men and 25% of the women are estrogen-driven, and that 30% of the men and 25% of the women are testosterone-driven (Derval 2010).

We will see in Chap. 2 that 'the organizational effects' of hormones have a direct impact on the 7 things that really matter in a job, in Chap. 3 how to use this knowledge to define our ideal career path, in Chap. 4 how to make the most of team work, and in Chap. 5 how to find the perfect balance.

References

Arnold AP, van Nas A, Lusis AJ (2009) Systems biology asks new questions about sex differences. Trends Endocrinol Metab 20(10):471–476

Cohen-Bendahan CC, van de Beek C, Berenbaum SA (2005) Prenatal sex hormone effects on child and adult sex-typed behavior: methods and findings. Neurosci Biobehav Rev 29(2):353–384

Derval D (2010) The right sensory mix: targeting consumer product development scientifically. Springer, Berlin

Fowden AL, Forhead AJ (2004) Endocrine mechanisms of intrauterine programming. Reproduction 127:515–526, 1 May 2004

Manning JT (2002) Digit ratio: a pointer to fertility, behavior, and health. Rutgers University Press, London

Marieb EN, Hoehn KN (2007) Human anatomy and physiology. Pearson, Benjamin Cummings

National Institutes of Health (2012a) How do genes direct the production of proteins? Genetics home reference. http://ghr.nlm.nih.gov/handbook/howgeneswork/makingprotein. Accessed 2012

National Institutes of Health (2012b) What are single nucleotide polymorphisms (SNPs)? Genetics home reference. http://ghr.nlm.nih.gov/handbook/genomicresearch/snp. Accessed 2012

National Institutes of Health (2012c) Can genes be turned on and off in cells? Genetics home reference. http://ghr.nlm.nih.gov/handbook/howgeneswork/geneonoff. Accessed 2012

Pinel JPJ (2007) Basics of biopsychology. Allyn & Bacon, Boston

Roughgarden J (2004) Evolution's rainbow: diversity, gender, and sexuality in nature and people. University of California Press, California

Varrone A (2011) Validation of a New Tracer in Humans. Route to Clinical Research for New PET Radioligands, Annual Congress of the European Association of Nuclear Medicine, Birmingham

Yoshiura K, Kinoshita A, Ishida T, Ninokata A, Ishikawa T, Kaname T, Bannai M, Tokunaga K, Sonoda S, Komaki R, Ihara M, Saenko VA, Alipov GK, Sekine I, Komatsu K, Takahashi H, Nakashima M, Sosonkina N, Mapendano CK, Ghadami M, Nomura M, Liang DS, Miwa N, Kim DK, Garidkhuu A, Natsume N, Ohta T, Tomita H, Kaneko A, Kikuchi M, Russomando G, Hirayama K, Ishibashi M, Takahashi A, Saitou N, Murray JC, Saito S, Nakamura Y, Niikawa N (2006) A SNP in the ABCC11 gene is the determinant of human earwax type. Nat Genet 38(3):324–330

Chapter 2
Unlock your Hormonal Quotient® (HQ)

> *Personality starts where comparison ends.*
> Karl Lagerfeld, Fashion Designer

In this chapter, we study the dimensions important to consider when evaluating a career path and how hormones influence each of them. We review main biomarkers like the shape of our face and the digit ratio, and what they reveal about our personality. We also explain how to measure the Hormonal Quotient® (HQ) and share the typical traits for each of the eight Hormonal Quotient® (HQ) profiles.

Seven Things That Really Matter in a Job

Finding a job is hard enough, so let's make sure it fits our profile. Most evaluation tools are based on professional competencies. But being good at numbers doesn't mean we will enjoy being an accountant. Why do some managers change position every year and others stay in the same job for at least 5 years? Why do some prefer dealing with products rather than with persons? Why are some people more competitive? Why are some managers obsessed by status when others are happy with a positive feedback? Why do some people appreciate detailed guidelines while others like freestyle? Why are some managers better at creating in teams and others on their own? There's more to a job than the actual tasks we perform. What really matters in a job is in fact related to the following critical dimensions: *movement, interaction, competition, recognition, structure,* and *collaboration.* And *prenatal hormones* have their say.

Movement: Sensation or Routine?

Movement is a critical dimension when considering a job as it can mean working outdoors, traveling, changing projects on a regular basis, or moving to another country. Amazingly, these aspects of our personality are all dictated by DRD4. No this is not a new Star Wars character, but a bunch of dopamine receptors (DR) of type

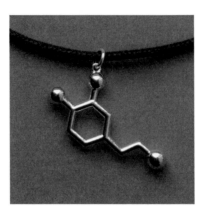

Fig. 2.1 Dopamine necklace by Raven Hanna www.madewithmolecules.com

D4, located in our body and brain. Dopamine (Fig. 2.1) is a critical messenger in reward-seeking behaviors, appetite, cognition, sociability, and movement.

And the DRD4 come in many different Single Nucleotide Polymorphisms (SNPs)—85 have been identified so far. 90% of the population present a 2R, 4R, or 7R polymorphism (EBI 2012).

Interestingly, the 7R polymorphism is linked with financial risk taking, migration, and food disorders. So basically people equipped with this 7R polymorphism in the DRD4 need to gamble or to relocate, if they want to stay in shape. In the working place, taking financial risks leads often to frequent job changes, taking on risky projects, accepting a larger variable part in our salary, or becoming our own boss. The classical day of work is therefore tense and full of surprises. The word "routine" is not in the vocabulary (Dreber et al. 2009).

Both 7R and 2R are linked with novelty-seeking traits and people presenting this SNP tend to be pioneers, impulsive, and over-the-top. 4R people tend to be more calm and measured. The 2R polymorphism is more represented among Asians and the 7R among Americans (Matthews and Butler 2011).

Recent research established a link between the 7R allele and Zuckerman Sensation Seeking Scale, but also with face masculinity, and circulating levels of testosterone (Campbell et al. 2010).

High sensation seekers tend to go for scientific, artistic, or social careers, becoming for instance a social worker, or a researcher. Individuals can express their sensation seeking in different ways (Zuckerman 1994). Men get involved in illegal car races or extreme sports. In fact, they don't perceive those activities as risky and their nervous system doesn't even generate the related fear, anxiety, or stress signals—a bit like orange-blue lizards (Roberti 2004).

We will see in Chap. 4 that some leaders, like orange-blue tree lizards, do not perceive stress signals related to risks and therefore understimate the likelihood of an event or its impact.

The Zuckerman Sensation Seeking Scale distinguishes four types of sensations: *thrill and adventure seeking*, like in scuba diving; *experience seeking*, like in going to metal concerts or visiting new countries; *disinhibition*, like in smoking marijuana and organizing orgies; and *boredom susceptibility*, like in having to constantly change TV channels. I would be more into experience seeking, what about you? (Zuckerman 1994).

Interaction: Object or Person?

What's wrong with colleagues coming to our desk or calling to discuss the email we just sent? Why are they so eager to interact? If like me you hate to work while being observed, avoid making phone calls, limit interactions and eye-contact, welcome to the world of social phobia. I score 67 (33 for fear and 34 for avoidance) on the Liebowitz Social Anxiety Scale Test. What about you? (Heimberg et al. 1999).

Social avoidance is connected with a low binding potential on Dopamine D2 receptors (DRD2). So is drug abuse (Schneier et al. 2000).

To overcome my fear of face-to-face interactions, I used to have a couple of drinks at networking events and tended to smoke a cigarette before each important phone call. Funnily enough, speaking in front of large audiences or complaining at a call center was never a problem (must be linked to another receptor!). I stopped those drugs 5 years ago as soon as I found the right career path for myself, with the right nature and frequency of interactions.

Social behavior is also linked to arginine vasopressin (AVPR1A) and oxytocin (OXTR) receptors. The expression and distribution of these peptide hormones depend on sex hormone: oxytocin (Fig. 2.2) is estrogen-dependent and vasopressin is testosterone-dependent.

Recent intranasal administration of oxytocin proved to increase the duration of eye-contact, and improve the ability to interpret face expressions. Also in different game plays, it made people trust more unknown business partners by reducing fear processing in different regions including the amygdala. The GG allele seems more prosocial than the AA and AG alleles, and people suffering from autism have 20% less oxytocin receptors. Before considering oxytocin nasal spray, you have to know that the effects are not all positive: higher levels of oxytocin generate envy.

Elevated expression of vasopressin is associated with anxiety and arousal. A SNP with a longer RS3 is linked with the age of first sexual intercourse in both males and females. Oxytocin and vasopressin are critical for remembering people (Ebstein et al. 2009).

The way social support helps us overcome stress is also linked to oxytocin. So having supportive friends is not always valued as it should (Mehta and Josephs 2010).

So the ideal working environment, open space or work from home, and ideal activities, involving face to face meetings or sitting in front of a computer, are determined by the density and shape of our dopamine D2, oxytocin, and vasopressin receptors. A library rat will avoid eye-contact while a sales representative can keep eye contact for minutes without even blinking, a bit like a chicken hypnotizer. Eye contact helps also apprehend face expressions and better manage social interactions.

Fig. 2.2 Oxytocin necklace by Raven Hanna www.madewithmolecules.com

Of course we can train by watching all the seasons of "Lie to Me" where a new generation of investigators use the fact that the 7 face expressions—anger, sadness, happiness, contempt, disgust, fear, and surprise—are innate and universal to spot criminals. It might be good to know though that this ability to understand other people just by looking at them is also part of our hormonal heritage. The estrogen receptors located in our retina will under the influence of prenatal hormones specialize each of us more on the M-pathway associated with motion or on the P-pathway associated with shapes and colors. This explains that people with a testosterone-driven Hormonal Quotient® (HQ) are more attracted by objects, especially if on wheels, whereas people with an estrogen-driven Hormonal Quotient® (HQ) are more attracted by shapes and colors, and are better at decrypting face expressions (Alexander 2003).

This specialization drives the type of toys we prefer. I was around 5 (Diana speaking), with a developed M-Pathway and at school we were celebrating Christmas. I couldn't believe my eyes when I saw this mountain of *racing cars* in the class room and was heading in their direction to grab my box when the teacher, a bit surprised, poked me and showed me where my gift actually was : she pointed in the direction of another table full of *baby-dolls*! What is fascinating is that I didn't even see the dolls, as my eyes were so fully captivated by the cars. Later in my career I was offered a job by two leading and innovative companies at the same time (a dream nowadays!), L'Oréal and Michelin: guess which company I picked?

The industry we work in, but also the working environment and type of activity is determined by our hormone receptors.

Can Your Neocortex Cope with your Facebook Friends?
The size of the neocortex is proportional to the number of connections apes have in their clique (Kudo and Dunbar 2001). Terrestrial individuals have a bigger network than arboreal individuals. So before networking extensively, you may want to check if your neocortex is able to manage your 500+ Facebook connections.

Fig. 2.3 Serotonin necklace by Raven Hanna www.madewithmolecules.com

Competition: Fight or Flight?

What makes a pitbull so scary? I mean apart from his owner. It's the fact that he will fight until he dies or at least he looks like it. When dogs bite, they do it for two main reasons: to defend their territory or to dominate. English springer spaniels, doberman pinschers, and toy poodles are often referred to clinical animal behaviorists for dominance aggression and German shepherds for protective aggression issues. With 65% of the biting victims being the actual owner, you might want to check carefully your dog's breed! (Houpt 2007).

In the workplace, it's a bit the same: some colleagues are trying to steal your project or role while others are doing their best to become your boss. Competition is in our nature: the *Fight or Flight* instinct. In many animal species, some individuals prefer to avoid confrontation and just run away, while others fight. Among the fighters some are clearly more resistant to defeat and ready for the next battle, think of our pitbull. To avoid constant fights, the function of our appearance—from pilo-erection in cats, to clothing, car, or body-building in humans—voice and eye contact is to impress the opponent to reach an agreement without even fighting.

In animals as well as in humans, hormones like serotonin (Fig. 2.3, also called hydroxytryptamine or HT), epinephrine, and norepinephrine seem largely involved. Animals with fewer 1-beta receptors for serotonin (HTR1B) show a greater aggressivity towards strange animals entering their territory, and also a higher sensitivity to pain, which could explain their defensive reaction (Houpt 2007).

HTR1B receptors are also associated with thermoregulation, appetite control, sexual behavior, and anxiety, as we will see in Chap. 5 (Weizmann Institute of Science 2012).

Among patients suffering from Alzheimer's disease, an aggressive behavior has been linked to a higher density of A (alpha) 2 adrenergic receptors (ADR). The ADRA2 receptors, like all the adrenergic receptors, bind with epinephrine and norepinephrine hormones. You are familiar with those as they are sometimes also called adrenaline and noradrenaline (all these different terms just to confuse us, for sure) (Russo-Neustadt and Cotman 1997).

In case of danger, the level of epinephrine and norepinephrine rises. The receptors ADRA2 and ADRA1 make sure the body is focused by generating vasoconstriction signals. We become tensed—if we are a cat or particularly hairy, a pilo-erection occurs—and second-priority activities like digestion are frozen in order to use all the resources to properly respond to the danger. When things are back to normal, our adrenergic beta receptors ADRB2 and ADRB1 ensure a vasodilation: our blood vessels are decontracted (NIH 2012).

I learned this vasoconstriction/vasodilation process the hard way the other day in front of the television. I was literally attacked on my sofa. Not exactly me, but a character in an episode of CSI. The aggression elevated my levels of norepinephrine, causing an asthma crisis. I grabbed the medicine prescribed for my asthma, Foradil, and for the first time read and understood its effect. It's a LABA: long-acting beta2-adrenergic agonist. Basically the drug activates the ADRB2 receptors responsible for the vasodilation so that I can come back to a relaxed stage quicker. It worked well. Writing this section of the book, I freaked out a bit when I found out that "Long-acting beta2-adrenergic agonists (LABA), such as formoterol the active ingredient in Foradil AEROLIZER, increase the risk of asthma-related death" (drugs.com 2012).

Great! Usually we can expect some side-effects with drugs but death sounds a bit extreme. Luckily, since I avoid watching violent programs like the daily news, sports, and action movies, I control my asthma much better.

Pharmacological tests established that depending on our ethnicity, our ADRB2 receptors are more or less sensitive. For instance, Chinese were found to be much more sensitive than afro-americans, which means that they can calm down easier (Mills et al. 1995).

Fight or Flight reactions occur even in front of the television, so imagine what happens in the work space—which is often a live soap opera. During a meeting, some will *fight* holding the marker as a weapon and imposing their view on the flipchart, while others will choose *flight* by reading their emails on their laptop.

The level of testosterone is also a predictor of our ability to fight again. After a defeat, 70% of the people with a high level of testosterone decided to fight again, against 22% among people with a lower testosterone level.

Also, in men with a high level of testosterone, the level of cortisol acted as a predictor in the *fight* or *flight* decision: low cortisol men chose to rechallenge their opponent after defeat, whereas high cortisol men chose to avoid a second competition. Testosterone seems to regulate our drive for status and cortisol our social approach or avoidance (Mehta and Josephs 2010).

Our propensity to fight or become upset as well as the time needed to calm down is directly linked to our adrenergic receptors, testosterone, and cortisol levels.

> **Brain Specialization**
> Most of the beasts, like birds, frogs, and mammals, manage their business in a very optimized way: their left brain hemisphere focuses on routine activities like eating while their right brain hemisphere is ready to face emergency situations like an attack (Vallortigara and Rogers 2005).

Recognition: Status or Feedback?

We are all looking for a certain form of recognition. Whether it translates into the number of followers on Twitter (by the way you can follow me @profdianaderval), a bonus, or a standing ovation.

Monkeys impress their peers by the number of fights they win. This helps them get more attention and dates. Research has established that monkeys looking for status have a higher binding potential on their dopamine receptors DRD2 and DRD3.

In the same way, research on human healthy volunteers revealed a correlation between status and dopamine receptors: people looking for social status have a higher binding potential on DRD2 and DRD3 (Martinez et al. 2010).

In birds, singing is the ultimate seduction tool—so singing is a bit like fighting. If dopamine motivates the bird to sing, the release of opioids is what makes singing rewarding and encourages the bird to sing again, and again, and sometimes too much, and too close to our ears, early in the morning (Riters 2010).

Our opioid receptors can be activated by many activities and substances. Sugar intake for instance has a direct impact on dopamine and opioid receptors and can be addictive for some people (Avena et al. 2008).

Each type of opioid receptor—OPRM1(mu), OPRK1(kappa), and OPRD1 (delta)—binds better with certain substances. For instance OPRM1 has a higher affinity with morphine, sugar, and b-endorphin (for endogenous morphine), released while doing exercise (Chahl 1996).

Opioid receptors release an analgesia when binding with an opioid. This has a "feel good" effect and can also relieve pain. Trying to explain why pain killers had more effect on certain individuals, researchers identified a direct connection between the estrogen receptors and the mu and kappa receptors. Among women, the sensitivity of these mu opioid receptors even follows the menstrual cycle explaining why pain killers activating these particular nociceptors (or pain receptors) are more effective during estrogen peaks. This confirms again the organizational and activational role of sex hormones on our overall reward system (Palmeira et al. 2011).

Our propensity to feel good after an action has a great influence on our learning experience and behavior. Typically being impulsive underlies not being able to cope with a delayed gratification and is activated by OPRM1 and deactivated by OPRD1 (Olmstead et al. 2009).

If we go back to our analysis of movement, impulsive people tend to have a 7R or a 2R allele on the dopamine D4 receptor and more mu opioid receptors while patient people tend to have a 4R allele on DRD4 and a higher density of delta opioid receptors.

Research also shows that basal testosterone is a biological marker of status-seeking in both men and women (Mehta and Josephs 2010). Hormones have a direct impact on our opioid receptors and on the type of reward we are looking for.

Structure: Analogical or Sequential?

My desk is a mess. This always caused me trouble while working in companies with a "clean desk" policy. The main reason for it, besides me being lazy, is that I never know how to classify documents. Shall I order them by project name, by client, by nature of project, by year?

I realized that my desk might actually be the tip of the iceberg. During meetings or *brainstorming*, I usually come up with ideas that others perceive as not related to the topic we discuss (ouch) or when they sound linked, then not in a *logical* way. To me these *crazy* or *intuitive* ideas are perfectly logical even if based on *unexpected* or *far-fetched* associations.

A recent publication saved me from group psychotherapy. People like me who think outside the box actually do not have a box at all: that's where the untidiness comes from! A low density of dopamine D2 receptors in the thalamus is associated with creative thinking and the ability to form uncommon solutions. This lack of dopamine D2 receptors by decreasing the information regulation and filtering, enables us to process multiple stimuli across a wider range of associations, and facilitates the switch between different idea combinations (Manzano et al. 2010).

This type of thinking is often referred to as reasoning by analogy. When mastered, analogy is a powerful path towards breakthrough innovation. The NASA engineer David Crocker had for instance an *aha-moment* under the shower, in his hotel room. At that time the Hubble mission started to sound like a disaster as the lens of the telescope was not properly adjusted before launch, resulting in a blurred space vision. As he was adjusting the shower head to his height, he designed by analogy the COSTAR system (Corrective Optics Space Telescope Axial Replacement) that saved the day by correcting the Hubble lenses while in orbit (Holyoak and Thagard 1995).

If you do not have a PET scan on hand to count DRD2 in the thalamus, you might as well want to try following test to identify the way of thinking of your colleagues and to adapt the way you present your ideas.

Here's a bear, a monkey, a banana (Fig. 2.4): how would you split them into two teams?

Seven Things That Really Matter in a Job   19

Fig. 2.4 A bear, a monkey, and a banana

Fig. 2.5 Sequential organization

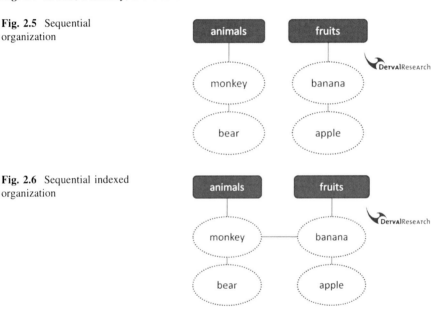

Fig. 2.6 Sequential indexed organization

(a) Bear and Monkey | Banana.
People grouping without any hesitation the bear with the monkey, because these two are animals, and fruits on the other side tend to have a higher density of dopamine D2 receptors. They store information into categories or boxes, and think in a sequential or linear way (Fig. 2.5). Novelty is disruptive as it requires sometimes the creation of new categories. On the other hand, analytical thinking is fast and effective as all the needed data is already grouped.

(b) Monkey and Bear | Banana or Monkey and Banana | Bear.
People hesitating between grouping the monkey with the bear or with the banana tend to have an average density of dopamine D2 receptors. They store information into categories or boxes, but at the same time create an index between other associations (Fig. 2.6). They can therefore think both in a

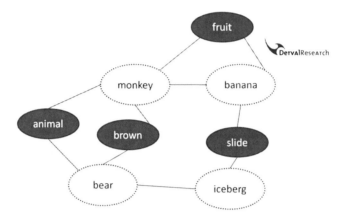

Fig. 2.7 Indexed organization

sequential and in an analogical way. They are key in cross-functional teams in order to translate communications between very sequential and very analogical people who usually do not understand each other.

(c) Monkey and Banana | Bear.

People grouping without any hesitation the monkey with the banana, because one eats the other, tend to have a lower density of dopamine D2 receptors. They store information together with their context and attributes (Fig. 2.7). For instance animal is an attribute at the same level than the color or the diet. Each attribute is an index and is linked to the others, which helps retrieve the data from many angles and together with the context. People can easily compare situations and come to new ideas by analogy.

> **Background Associations**
> When we look like we are doing nothing, our brain might be busy making associations between data and objects we memorized in order to generate predictions (Bar 2007).

Collaboration: Input or Co-Creation?

Some of us enjoy teamwork and some work best in teams of one. Of course, any type of work requires gathering feedback and confronting ideas but there's quite a difference between getting others' input and then making a decision or shaping a decision together with others, in a co-creation mode.

If you remember from previous pages, a low binding potential of DRD2 is also linked to social avoidance. Quite challenging to be creative in teams if creative people avoid team work, huh? One way to solve this dilemma is to arrange one to one interviews with people having a low density of dopamine D2 receptors rather than getting together people having a high density of DRD2.

Once again testosterone and estrogen have their say. Twenty people's reaction was compared in a cooperative and in a competitive context. They were split by teams of two and had to answer individually and without consultation within the team 15 questions from the former Graduate Record Exam (GRE). In the competitive context, the best of each pair would win $25. In the cooperative context, if the team score was higher than the score of the following team, both participants would win $25 each. The finding was that people with a higher basal testosterone level (measured via the saliva) performed well in the competitive context and poorly in the cooperative context (Mehta and Josephs 2010).

In birds, high testosterone birds are singing and defending the nest while low testosterone birds are busy with flock guidance (Heimovics et al. 2009).

Testosterone-driven people are not the best at collaborating unless the outcome of the team work can serve their goals such as gaining more status (Mehta and Josephs 2010).

Find Your Own Path

When it's possible, why not have fun at work? In our constant quest for pleasure, we might want to make each minute of our day enjoyable.

Our opioid receptors might respond to different stimuli than our colleagues. Some of us enjoy spending time in teams, others would kill for a private office space. Some need to be followed on Twitter by an "entourage" of supporters, while others feel enough recognition driving the latest BMW.

The number of our oxytocin and vasopressin receptors is key for our interpersonal skills. Adrenergic and serotonin receptors condition the way we fight-or-flight. Our basal testosterone has its say on the way we collaborate in teams. The binding potential of dopamine receptors is involved in our need for movement, interaction, structure, and recognition. At the center of this motivation and reward system, our opioid receptors pull the strings (Fig. 2.8).

As we saw in Chap. 1, our hormones receptors are shaped by prenatal hormones. This means that the number, affinity, and type of our opioid receptors—remember the mu, delta, and other kappa—and consequently what motivates us in life is also determined by our Hormonal Quotient® (HQ).

Recent research confirmed that our level of aggression (Benderlioglu and Nelson 2004), anxiety (De Bruin et al. 2006; Evardone and Alexander 2009), hyperactivity (Williams et al. 2003), willingness to cooperate (Millet and Dewitte 2006), and sensation seeking (Campbell et al. 2010) are all influenced by prenatal hormones. Testosterone-driven people are for instance more likely to be hyperactive. This would be a gift in a fast-moving company and a curse in an administrative context.

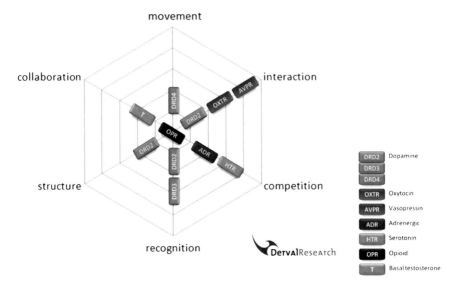

Fig. 2.8 7 Things that matter in a job

The best way to find your own path is to measure your Hormonal Quotient® (HQ)!

Biomarkers and Hormonal Quotient® (HQ)

Our Hormonal Quotient® (HQ) can be reliably inferred by the shape of our face (Burriss et al. 2007; Neave et al. 2003; Schaefer et al. 2005), and the relative length of our index and ring fingers (called digit ratio).

What You See Is What You Get

Many physical and personality traits reveal our hormonal mark up.

Testosterone-driven people for instance tend to have a wider neck (Fink et al. 2003), a lower voice (Ferdenzi et al. 2011), and a handwriting difficult to decode, especially in women (Beech and Mackintosh 2005).

Most left-handed people were also influenced by prenatal testosterone (Stoyanov et al. 2009; Fink et al. 2004).

Estrogen-driven people tend to articulate better, and have a richer vocabulary (Beech and Beauvois 2006; Albores-Gallo et al. 2009). Estrogen men tend to have a higher body mass index (Fink et al. 2003).

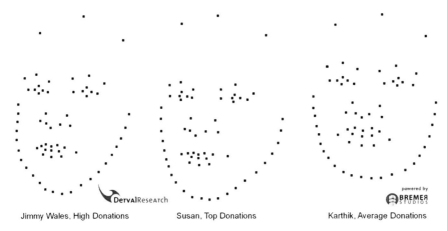

Fig. 2.9 Wikipedia fund raising campaign faces

This could explain why *physiognomy*—nothing to do with gnomes, just the assessment of the personality by the analysis of the morphology and especially the face—is often used by recruiters and investors. Let's see how bankable you are.

How Bankable Are You?

Wikipedia just ran a successful fund raising campaign to finance the maintenance and development of the 3 million articles posted. Wikipedia users around the globe joined forces, and resources, and donated this year $20 million.

When visiting their favorite encyclopedia, they were exposed to a banner with a picture of Wikipedia founder Jimmy Wales. Once they clicked on it, a message was explaining the objective of the campaign and inviting them to donate via a Paypal link. To engage their users, Wikipedia alternated the picture of Jimmy Wales with pictures of local Wikipedians, the people who populate the articles of the encyclopedia on a voluntarily basis.

Interestingly, a certain Susan, researcher in mollusks, generated more clicks and donations than all the other Wikipedians together and even more than the founder Jimmy Wales himself (Wikimedia Foundation 2012).

What made people want to help Susan? Her serious face expression, with respectable grey hair. Oxytocin in her look? Or maybe the shape of her face? We will see in Chap. 5 that the shape of the face is key in dating, so why not in donating? If we compare the faces of Susan, Jimmy Wales, and Karthik, a Wikipedian who generated less donations—using the 64 reference pixels of the face to facilitate comparison (Fink et al. 2005)—we see that Jimmy and Susan have longer faces (Fig. 2.9).

We decided to have a closer look at some successful business leaders like Steve Jobs and Mark Zuckerberg. They both have long faces, and their wealth is

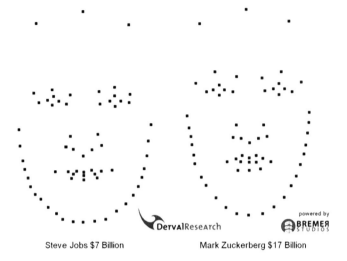

Fig. 2.10 Successful leaders faces

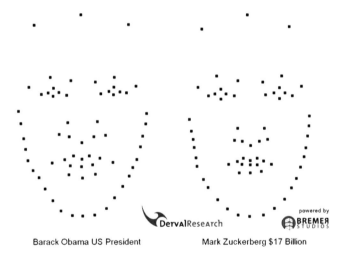

Fig. 2.11 Next US president?

estimated respectively at $7 billion and $17 billion (Fig. 2.10). Steve Job's face shape is even very similar to Susan's one. The length of the face seems at least to be an indicator of wealth. Facebook founder, Mark Zuckerberg has a very long chin that reminded us of someone else.

We decided to broaden our investigation to some popular faces in politics.

Barack Obama, the U.S. president, also has a long face, a long chin, and even a thinner chin than Facebook's founder (Fig. 2.11). Based on physiognomy, Mark Zuckerberg could well be a future U.S. president.

Research on face shape and hormones, indicates that the lengthening of the face and the lower face is related to high levels of testosterone (Fink et al. 2005).

Barack Obama is left-handed (like most U.S. presidents) (Hutchison 2011) and has a low digit ratio (we will see what this means shortly), the influence of testosterone seems confirmed here. Funnily enough, Mark Zuckerberg is left-handed as well. And Steve Jobs declared himself ambidextrous.

Talking about U.S. presidents, billionaire Donald Trump got involved in the U.S. presidential run. Let's have a closer look at his profile.

Real estate mogul Donald Trump is worth $3 billion, he chairs the reality tv-show "The Apprentice" and he wrote several best-selling business books, such as "The Art of the Deal" and "Think Like a Billionaire" (Trump 2012).

Descended from German immigrants—his father was already a big name in real estate (Blair 2001)—Donald is clearly into movement and has probably a high dopamine receptors D4 count as seen in this Chapter.

Where his father built a $200 million company developing affordable housing, Donald managed to multiply it by ten focusing on luxury buildings (A&E Television Networks 2012).

Status is clearly important for Trump: he put his name on 36 skyscrapers (Skyscraper Source Media 2012), got a trademark on the name Trump (which brands better than the original family name Drumpf) (Blair 2001), co-owns the Miss Universe Pageant, was married to two models and an actress (A&E Television Networks 2012), and even donates money for children who need reconstructive surgery (Trump 2012).

Donald Trump's focus on status and recognition would indicate a high count in dopamine receptors D2 and D3.

Donald's physiognomy, especially his signature hairstyle, have been the subject of much research. Time magazine actually published a manual to recreate his coiffure (Time inc. 2012).

The man may have funny hair, but he has a bankable face: His wealth multiplied from $200 million to $1.7 billion, melted down to a debt of $900 million, then back growing up to $3 billion. Even during the crisis and with a colossal debt, he convinced banks to invest in his high-profile real estate projects, and obviously it paid off well (O'Brien 2012).

With his strong need for status and movement, no wonder Donald considered running for president (CNN Political Unit 2011).

Donald Trump doesn't have the longest face, but his profile presents another strong attribute linked to prenatal testosterone: the forward growth of the bones of the eyebrows ridge (Fig. 2.12). We decided to call it the "Klingon ridge" (Star Trek fans will appreciate this). It is clearly visible on Donald Trump's pixelized profile. His cheeks are wide, also a sign of high levels of testosterone.

So for your CV picture, you might want to consider some serious Photoshopping in order to make your face, or chin longer, alternatively to add a "Klingon

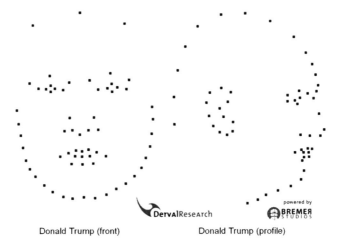

Fig. 2.12 Donald Trump

ridge." Smiling would also work—in the Wikipedia campaign, the same woman smiling generated 5% more donations. If it works for donators, voters, and investors, it will surely work for recruiters.

Our Future Is in Our Hands

I should say, in our fingers. Although testosterone and dihydroxytestosterone (DHT)—a high binding testosterone responsible for very masculine traits like baldness—affect the development of the epidermis and dermis in fingers (Jamison et al. 1993) there is too little research for now to enable us to read the future in our hand palms (Manning et al. 2000).

However, it has been proven that the relative length of our index and ring finger is a biomarker for prenatal hormones (Fig. 2.13). And here is how it works: it is all about receptors again. The ring finger has more androgen receptors (AR) and estrogen receptors-α (ER) than the index. Inactivation of AR decreases the growth of the ring finger, leading to a higher 2D:4D ratio, whereas inactivation of ER-α increases the growth of the ring finger, leading to a lower 2D:4D ratio (Zheng and Cohn 2011).

The ratio between the index and ring finger is determined when we are a fetus and remains the same. It is considered as a reliable biological marker of the 'organizational effects' of hormones. Our digit ratio or hormonal fingerprint can be

Biomarkers and Hormonal Quotient® (HQ)

Fig. 2.13 Digit ratio measurement

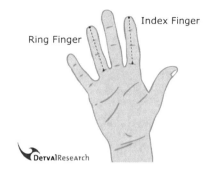

Fig. 2.14 HQ wheel

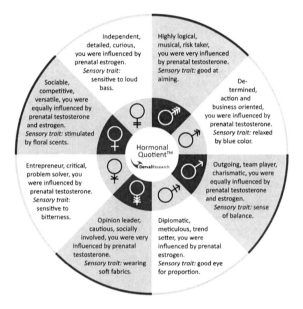

measured by comparing the length between the index and the ring finger: a shorter index is a sign of higher exposure to prenatal testosterone (Manning 2002).

Among the several ways of measuring prenatal hormones, the digit ratio technique is the least invasive and the handiest.

Measure Your Hormonal Quotient® (HQ)

Based on measurements performed on 3,500 people in over 25 countries, we were able to identify the main personality traits for each of the 8 Hormonal Quotient® (HQ) profiles (Fig. 2.14).

Fig. 2.15 Diana Derval and Johan Bremer measuring celebrities handcasts

We even measured with a digital vernier calliper celebrities' handcasts from the Grauman Chinese Theater in Hollywood to Wembley Stadion's Hall of Fame in London (Fig. 2.15). And by the way, it is confirmed: Donald Trump is testosterone-driven!

And here are the findings for men (Fig. 2.16) and for women (Fig. 2.17). You can easily check your HQ at www.derval-research.com!

Men with a very testosterone-driven Hormonal Quotient® (HQ) are highly logical, musical, and risk takers.

Famous men with a very testosterone Hormonal Quotient® (HQ): Barack Obama, Daniel Radcliffe (famous from his role as Harry Potter), Danny Glover, Leonard Nimoy, Steve McQueen, John Woo, Will Smith, William Shatner, Steve Jobs.

Men with a testosterone Hormonal Quotient® (HQ) are determined, action and business oriented.

Famous men with a testosterone Hormonal Quotient® (HQ): Donald Trump, Bruce Willis, Cliff Richard, Yul Brynner, Eddie Murphy, Anthony Hopkins, Bing Crosby, Jack Nicholson, Rick Parfitt (Status Quo), Michael Douglas, Johnny Depp, Lionel Richie, Stuart Copeland (The Police), Steven Seagal, Morgan Freeman, Nicolas Cage, Richard Gere, Samuel L Jackson, Stephen Hendry (star snooker player), Steven Spielberg, Sting, Warren Beatty, Tom Hanks, Arnold Schwarzenegger.

Men with a balanced Hormonal Quotient® (HQ) are outgoing, team players, and charismatic.

Famous men with a balanced Hormonal Quotient® (HQ): George Clooney, Matt Damon, Michael Jackson, Bryan Adams, Clark Gable, George Michael, John Travolta, Gregory Peck, Andy Summers (The Police), Paul 'Showtime' Gaffney (from the Harlem Globetrotters), Rupert Grint, George Lucas, Ron Howard.

Men with an estrogen Hormonal Quotient® (HQ) are diplomatic, meticulous, and trend setter.
Famous men with a balanced Hormonal Quotient® (HQ): Brad Pitt, Harrison Ford, Mel Gibson, Robin Williams, Cecil B Demille, Paul Newman, Sylvester Stallone, Clint Eastwood.

Fig. 2.16 Hormonal Quotient® (HQ) in men

Fig. 2.17 Hormonal Quotient® (HQ) in women

References

Albores-Gallo L, Fernández-Guasti A, Hernández-Guzmán L, List-Hilton C (2009) 2D:4D finger ratio and language development. Rev Neurol 48 (11):577–581 (PubMed PMID:19472155)

Alexander GM (2003) An evolutionary perspective of sex-typed toy preferences: pink, blue, and the brain. Arch Sex Behav 32(1):7–14

Avena NM, Rada P, Hoebel BG (2008) Evidence for sugar addiction: behavioral and neurochemical effects of intermittent, excessive sugar intake. Neurosci Biobehav Rev 32(1):20–39

A&E Television Networks (2012) Donald trump biography. http://www.biography.com/people/donald-trump-9511238?page=3. Accessed 2012

Bar M (2007) The proactive brain: using analogies and associations to generate predictions. Trends Cogn Sci 11(7):280–289

Beech JR, Mackintosh IC (2005) Do differences in sex hormones affect handwriting style? Evidence from digit ratio and sex role identity as determinants of the sex of handwriting. Pers Individ Differ 39(2):459–468

References

Beech JR, Beauvois MW (2006) Early experience of sex hormones as a predictor of reading, phonology, and auditory perception. Brain Lang 96(1):49–58 (PubMed PMID:16203032)

Benderlioglu Z, Nelson RJ (2004) Digit length ratios predict reactive aggression in women, but not in men. Horm Behav 46(5):558–564 (PubMed PMID:15555497)

Blair G (2001) The Trumps: three generations that built an empire. Simon & Schuster, New York

Burriss RP, Little AC, Nelson EC (2007) 2D:4D and sexually dimorphic facial characteristics. Arch Sex Behav 36(3):377–384 (PubMed PMID:17203400)

Campbell BC, Dreber A, Apicella CL, Eisenberg DT, Gray PB, Little AC, Garcia JR, Zamore RS, Lum JK (2010) Testosterone exposure, dopaminergic reward, and sensation-seeking in young men. Physiol Behav 99(4):451–456 (PubMed PMID:20026092)

Chahl LA (1996) Opioids—mechanisms of action. Aust Prescr 19:63–65

CNN Political Unit (2011) Trump not running for president. http://politicalticker.blogs.cnn.com/2011/05/16/breaking-trump-not-running-for-president/?hpt=T2. Accessed 2012

De Bruin EI, Verheij F, Wiegman T, Ferdinand RF (2006) Differences in finger length ratio between males with autism, pervasive developmental disorder-not otherwise specified, ADHD, and anxiety disorders. Dev Med Child Neurol 48(12):962–965

Dreber A, Apicella CL, Eisenberg DTA, Garcia JR, Zamore RS, Lum JK, Campbell B (2009) The 7R polymorphism in the dopamine receptor D_4 gene (DRD4) is associated with financial risk taking in men. Evolut Hum Behav 30:85–92

Drugs.com (2012) Foradil Official FDA information, side effects and uses. http://www.drugs.com/pro/foradil.html. Accessed 2012

EBI (2012) Gene and protein summary: DRD4. http://www.ebi.ac.uk/s4/summary/molecular?term=DRD4&classification=9606&tid=nameOrgENSMUSG00000025496. Accessed 2012

Ebstein RP, Israel S, Lerer E, Uzefovsky F, Shalev I, Gritsenko I, Riebold M, Salomon S, Yirmiya N (2009) Arginine vasopressin and oxytocin modulate human social behavior. Ann N Y Acad Sci 1167:87–102

Evardone M, Alexander GM (2009) Anxiety, sex-linked behaviors, and digit ratios (2D:4D). Arch Sex Behav 38(3):442–455 (PubMed PMID:17943431; PubMed Central PMCID: PMC2768336)

Ferdenzi C, Lemaître JF, Leongómez JD, Roberts SC (2011) Digit ratio (2D:4D) predicts facial, but not voice or body odour, attractiveness in men. Proc Biol Sci (PubMed PMID:21508034)

Fink B, Neave N, Manning JT (2003) Second to fourth digit ratio, body mass index, waist-to-hip ratio, and waist-to-chest ratio: their relationships in heterosexual men and women. Ann Hum Biol 30(6):728–738 (PubMed PMID:14675912)

Fink B, Manning JT, Neave N, Tan U (2004) Second to fourth digit ratio and hand skill in Austrian children. Biol Psychol 67(3):375–384 (PubMed PMID:15294393)

Fink B, Grammer K, Mitteroecker P, Gunz P, Schaefer K, Bookstein FL, Manning JT (2005) Second to fourth digit ratio and face shape. Proc Biol Sci 272(1576):1995–2001

Heimberg RG, Horner KJ, Juster HR, Safren SA, Brown EJ, Schneier FR, Liebowitz MR (1999) Psychometric properties of the liebowitz social anxiety scale. Psychol Med 29(1):199–212 (PubMed PMID:10077308)

Heimovics SA, Cornil CA, Ball GF, Riters LV (2009) D1-like dopamine receptor density in nuclei involved in social behavior correlates with song in a context-dependent fashion in male European starlings. Neuroscience 159(3):962–973 (Epub 2009 Jan 27)

Holyoak KJ, Thagard P (1995) Mental leaps: analogy in creative thought. The MIT Press, Boston

Houpt KA (2007) Genetics of canine behavior. Acta Vet. Brno 76:431–444

Hutchison C (2011) Left-handed legacy: demystifying the southpaw. ABC News Medical Unit: http://abcnews.go.com/blogs/health/2011/09/27/left-handed-legacy-demystifying-the-southpaw. Accessed 2012

Jamison CS, Meier RJ, Campbell BC (1993) Dermatoglyphic asymmetry and testosterone levels in normal males. Am J Phys Anthropol 90(2):185–198

Kudo H, Dunbar RIM (2001) Neocortex size and social network size in humans. Anim Behav 62:711–722

Manning JT, Barley L, Walton J, Lewis-Jones DI, Trivers RL, Singhe D, Thornhill R, Rohdeg P, Bereczkeih T, Henzii P (2000) The 2nd:4th digit ratio, sexual dimorphism, population differences, and reproductive success: evidence for sexually antagonistic genes? Evolut Hum Behav 21:163–183

Manning JT (2002) Digit Ratio: a pointer to fertility, behavior, and health. Rutgers University Press, London

Manzano Ö, Cervenka S, Karabanov A, Farde L, Ullén F (2010) Thinking outside a less intact box: thalamic dopamine D2 receptor densities Are negatively related to psychometric creativity in healthy individuals. PLoS ONE 5(5):e10670

Martinez D, Orlowska D, Narendran R, Slifstein M, Liu F, Kumar D, Broft A, Van Heertum R, Kleber HD (2010) Dopamine type 2/3 receptor availability in the striatum and social status in human volunteers. biological psychiatr 67(3):275–278

Matthews LJ, Butler PM (2011) Novelty-seeking DRD4 polymorphisms are associated with human migration distance out-of-Africa after controlling for neutral population gene structure. Am J Phys Anthropol 145(3):382–389

Mehta PH, Josephs RA (2010) Social endocrinology. In: Dunning D (ed) Social motiva-tion. Psychology Press, New York, pp 171–189

Millet K, Dewitte S (2006) Second to fourth digit ratio and cooperative behavior. Biol Psychol 71(1):111–115

Mills PJ, Dimsdale JE, Ziegler MG, Nelesen RA (1995) Racial differences in epinephrine and ß$_2$-adrenergic receptors. Hypertension 25:88–91

Neave N, Laing S, Fink B, Manning JT (2003) Second to fourth digit ratio, testosterone and perceived male dominance. Proc Biol Sci 270(1529):2167–2172 (PubMed PMID:14561281; PubMed Central PMCID: PMC1691489)

NIH (2012) Vasoconstriction. Medlineplus: http://www.nlm.nih.gov/medlineplus/ency/article/002338.htm. Accessed 2012

O'Brien TL (2012) What's he really Worth? *The New York Times*. http://www.nytimes.com/2005/10/23/business/yourmoney/23trump.html. Accessed 2012

Olmstead MC, Ouagazzal AM, Kieffer BL (2009) Mu and delta opioid receptors oppositely regulate motor impulsivity in the signaled nose poke task. PLoS ONE 4(2):e4410

Palmeira CCA, Ashmawi HA, Posso IP (2011) Sex and pain perception and analgesia. Rev Bras Anestesiol 61(6):814–828

Riters LV (2010) Evidence for opioid involvement in the motivation to sing. J Chem Neuroanat 39(2):141–150 (Epub 2009 Apr 5)

Roberti JW (2004) A review of behavioral and biological correlates of sensation seeking. J Res Pers 38(3):256–279

Russo-Neustadt A, Cotman CW (1997) Adrenergic receptors in alzheimer's disease brain: selective increases in the cerebella of aggressive patients. J Neurosci 17(14):5573–5580

Schaefer K, Fink B, Mitteroecker P, Neave N, Bookstein FL (2005) Visualizing facial shape regression upon 2nd to 4th digit ratio and testosterone. Coll Antropol 29(2):415–419

Schneier FR, Liebowitz MR, Abi-Dargham A, Zea-Ponce Y, Lin SH, Laruelle M (2000) Low dopamine D2 receptor binding potential in social phobia. Am J Psychiatry 157:457–459

Skyscraper Source Media (2012) SkyscraperPage.com. http://skyscraperpage.com/diagrams/?searchID=53545506. Accessed 2012

Stoyanov Z, Marinov M, Pashalieva I (2009) Finger length ratio (2D:4D) in left- and right-handed males. Int J Neurosci 119(7):1006–1013 (PubMed PMID:19466635)

Time Inc (2012) The secret to Donald Trump's hair. http://www.time.com/time/interactive/0,31813,2065167,00.html. Accessed 2012

Trump DJ (2012) Donald Trump biography. http://www.trump.com/Donald_J_Trump/Biography.asp. Accessed 2012

Vallortigara G, Rogers LJ (2005) Survival with an asymmetrical brain: advantages and disadvantages of cerebral lateralization. Behav Brain Sci 28(4):575–589 discussion 589–633

Weizmann Institute of Science (2012) HTR1B Gene. GeneCards: http://www.genecards.org/cgi-bin/carddisp.pl?gene=HTR1B&search=HTR1B. Accessed 2012

Wikimedia Foundation (2012) Fundraising 2011. Meta-Wiki: http://meta.wikimedia.org/wiki/Fundraising_2011, Accessed 2012

References

Williams JH, Greenhalgh KD, Manning JT (2003) Second to fourth finger ratio and possible precursors of developmental psychopathology in preschool children. Early Hum Dev 72(1):57–65

Zheng Z, Cohn MJ (2011) Developmental basis of sexually dimorphic digit ratios. PNAS 108(39):16289–16294

Zuckerman M (1994) Behavioral expressions and biosocial bases of sensation seeking. Cambridge University Press, Cambridge

Chapter 3
Find the Right Career Path

> *Be nice to nerds, chances are you'll end up working for one.*
> Bill Gates, Founder of Microsoft.

In this chapter, we analyze job fit from a physiological perspective. The ideal jobs (Fig. 3.1), as well as the career paths to avoid, are reviewed for each Hormonal Quotient® (HQ) profile. We also reveal the Hormonal Quotient® (HQ) of leaders—from firms including *Sofitel*, *Solvay*, *Philips*, and *L'Oréal* —and explain how they used their talent to build a successful career.

Physiology and Job Fit

Of course, we can potentially do many types of jobs but in order to last we need to fulfill the basic skills required.

Research conducted on 134 firefighters showed that individuals with a more balanced HQ were best suited for the job—always useful to be balanced when climbing a ladder. Firefighters too much on the testosterone side had lower service ranks (Voracek et al. 2010). Things are different in the world of rugby. Among 44 elite rugby players from the Ospreys Rugby Union Club, the most testosterone-driven ones were performing better in that occupation with a higher number of caps (Bennett et al. 2010). For those like me less familiar with rugby, this doesn't mean they are wearing several hats. When rugby started, players were not wearing the same shirt but the same cap. So being selected for a game meant getting a cap. Even in more quiet careers like in academia, hormones have a huge influence: men and women making careers as scientists are balanced, while men and women pursuing a path as social scientists are testosterone-driven (Brosnan 2006).

Hormonal Quotient® (HQ) and Occupation in Women

I noticed the link between hormones and career, when with DervalResearch we started segmenting consumers for companies like Sara Lee, or Philips in order to

Fig. 3.1 Is it the right career path?

develop the right products or promotion plans. People with the same Hormonal Quotient® (HQ) tended to have the same occupation.

Here are some measurements we performed on a sample of 67 female shoppers back in 2007 (Fig. 3.2). We were interviewing them about their product preferences and measuring their physiological traits. Our findings have since then been confirmed and enriched by tests done on 3,500 people.

Estrogen women were mainly civil servants, employees, or stay-at-home moms, with a structured environment. Balanced women were doctors, medical assistants, physical therapists, or teachers—more into interaction. Testosterone women were dancers, entrepreneurs, fashion designers, shop owners—more into creation. For very testosterone women hobbies were more important than their jobs as psychologist assistants, product managers, and some had the chance to make a living doing something they love like acting.

This is just a snapshot in a given shopping area at a given time and all the jobs women might have are not represented. What is fascinating though is to see that many women with a similar job also have the same HQ. For instance, call center agents and entrepreneurs were testosterone-driven, medical assistants balanced, and civil servants on the estrogen side. Very-testosterone driven women rather qualify themselves unemployed than stay at home.

Also, when we are limited in our career choices because of education or economic crisis, there is always one job that is less bad than another. So in the following sections we will not only review the ideal jobs but also the career paths to avoid.

We wanted to understand on a physiological level how prenatal hormones can shape our future career choices. Here are the findings on career choices and working environment.

Physiology and Job Fit

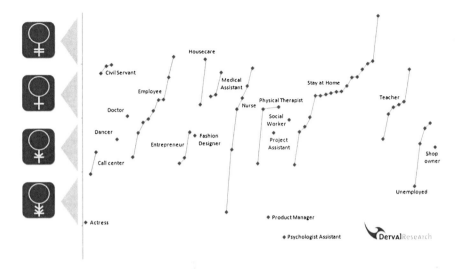

Fig. 3.2 Sample of Women Occupations per Hormonal Quotient®(HQ) *Source* Research conducted by DervalResearch on 67 Dutch female shoppers in 2007

Vision and Career

With the collaboration of Carl Zeiss opticians, we analyzed in Belgium the eye of 49 subjects, their job, hobbies, and Hormonal Quotient® (HQ).

You can interpret the pictures of the retina below taken with Carl Zeiss i.Profiler as a topographic map: the more you have mountains or valleys, the more you have imperfections on the eye called aberrations. In a fascinating way, people working with colors and shapes, such as web designers, florists, architects, interior designers had fewer aberrations in their eyes, see Fig. 3.4. This means that their view of colors and shapes is perfect. They can differentiate between many nuances of colors. All of them, men and women, had also an estrogen-driven Hormonal Quotient® (HQ).

Men and women with a testosterone-driven Hormonal Quotient® (HQ) presented many aberrations as seen on Fig. 3.3. This is not optimal for color discrimination but helps for perceiving depth and 3D, a bit like a camera creating a 3D effect by recording from different angles (Derval 2010a). Useful for other types of jobs, like a pilot, or a taxi driver, for instance.

Friendly Work Environment

The working environment is often as important as the job itself. We were wondering what makes it friendly and found the answer in our ears. Twenty-five

Fig. 3.3 Eye of testosterone-driven men and women (Carl Zeiss i.Profiler analysis)

Fig. 3.4 Eye an estrogen-driven men and women (Carl Zeiss i.Profiler analysis)

thousand hair cells, located in our inner ear, help us perceive sound. Each of their stereocilia captures and amplifies a certain frequency (Fig. 3.5).

Our inner-ear works like an amplifier. Depending on their Hormonal Quotient® (HQ), people amplify sound in very different ways. We observed for instance that Indians tend to add extra bass whereas Chinese don't like too much 'boom-boom' as they call it. Some individuals hear a same sound more than 4 times louder than others. Imagine how disruptive it must be for them to work in an open office!

Haircells and stereocilia generate their own and unique noise when amplifying sound. This noise is called otoacoustic emissions (OAE) and can be measured in the ear with a special reader. We measured these otoacoustic emissions on 16 healthy male volunteers, with a clinical OAE reader provided by Interacoustics, and asked the subjects questions about their favorite music, working configuration, and job and hobbies. We were able to group them according to their hearing sensitivity into 3 groups: non-amplifiers, medium-amplifiers, and super-amplifiers (Derval 2010a).

Hormonal Quotient® (HQ) and Ideal Job

Fig. 3.6 Meet Virginie Belliere-Baca, Solvay

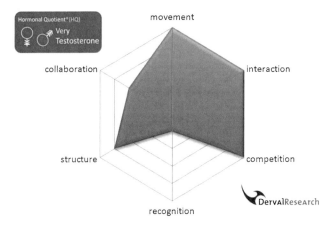

Fig. 3.7 Very Testosterone HQ and Career Profile *Source* Research conducted by DervalResearch on 400 managers and leaders, men and women, from over 20 countries enrolled in executive education programs

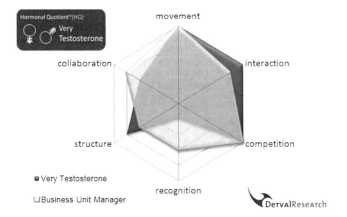

Fig. 3.8 Very Testosterone HQ and Business Unit Manager *Source* Research conducted by DervalResearch on 400 managers and leaders, men and women, from over 20 countries enrolled in executive education programs

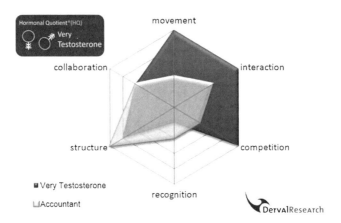

Fig. 3.9 Very Testosterone HQ and Accountant *Source* Research conducted by DervalResearch on 400 managers and leaders, men and women, from over 20 countries enrolled in executive education program

bored to death on the movement and competition side (Fig. 3.9). An option could be to create a firm in accounting and become its general manager.

Testosterone HQ: Product Manager versus Doctor

Testosterone-driven people tend to think outside the box—they must have a low binding potential on dopamine D2 receptors. They are therefore very good at tasks

Hormonal Quotient® (HQ) and Ideal Job

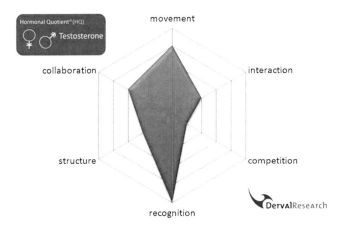

Fig. 3.10 Testosterone HQ and Career Profile *Source* Research conducted by DervalResearch on 400 managers and leaders, men and women, from over 20 countries enrolled in executive education programs

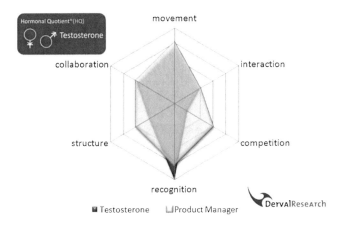

Fig. 3.11 Testosterone HQ and Product Manager *Source* Research conducted by DervalResearch on 400 managers and leaders, men and women, from over 20 countries enrolled in executive education programs

involving creative problem solving. They value the input of the team but like to make the final decision on their own (Fig. 3.10).

Top jobs: entrepreneur, finance director, IT director, management consultant, lead engineer, product manager, key account manager in automotive or telecoms, technical director, CEO, auditor, program manager, geoscientist.

Fig. 3.12 Meet Nick Goldie, Philips Consumer Lifestyle

A position as a product manager can suit testosterone-driven men and women as it is a perfect match on collaboration, movement, and interaction, and challenging on the competition part. If the industry and product is appealing and fast moving this could be the perfect match (Fig. 3.11). Like Nick Goldie, Finance Director at Philips Consumer Lifestyle France, leaders with a testosterone HQ tend to be more into collaboration than into competition: they put their team first (Fig. 3.12).

Jobs to avoid: lawyer, accountant, doctor.

Being a doctor involves way too much interactions and structure for testosterone-driven people (Fig. 3.13).

Hormonal Quotient® (HQ) and Ideal Job

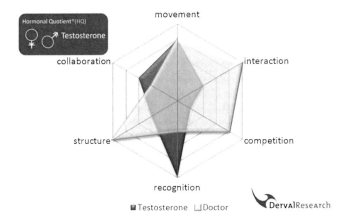

Fig. 3.13 Testosterone HQ and Doctor *Source* Research conducted by DervalResearch on 400 managers and leaders, men and women, from over 20 countries enrolled in executive education programs

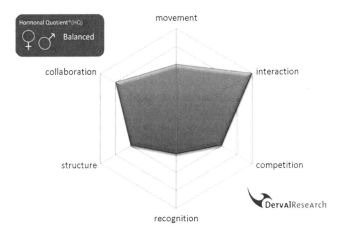

Fig. 3.14 Balanced HQ and Career Profile *Source* Research conducted by DervalResearch on 400 managers and leaders, men and women, from over 20 countries enrolled in executive education programs

Balanced HQ: Operations Director versus Surgeon

Balanced men and women are not lacking oxytocin receptors: they are very sociable (Fig. 3.14). They remember people's faces, the names of their friend's children, and favorite hobbies, which makes them perfect team players but also sales people (Ebstein et al. 2009).

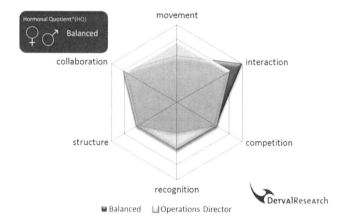

Fig. 3.15 Balanced HQ and Operations Director *Source* Research conducted by DervalResearch on 400 managers and leaders, men and women, from over 20 countries enrolled in executive education programs

Feeling the support of a team helps them overcome challenges and take some measured risks (Mehta and Josephs 2010).

Top jobs: Logistics director, operations director, retail director, area manager, industrial director, real estate manager, sales manager medical devices and chemicals, management consultant, head of sales in banking, international civil servant, nursing manager, marketing manager in pharmaceuticals, senior physiotherapist, project manager, ecommerce manager, MD.

Jobs that are ideal for balanced men and women involve a lot of interactions with peers and work in cross-functional teams. An operations director is a good example of a position with the right amount of novelty, structure, and competition (Fig. 3.15).

Jobs to avoid: surgeon, innovation manager.

Like most balanced men and women, Ariane Latreille, Retail Director at CCB L'Oréal, doesn't enjoy tasks involving too much structure, and too little competition (Fig. 3.16).

A job to avoid would be surgeon for instance. It requires at the same time the command of a certain number of routines, performed in teams, and the ability to make critical decisions in a split second (Fig. 3.17).

Hormonal Quotient® (HQ) and Ideal Job

Fig. 3.16 Meet Ariane Latreille, L'Oréal

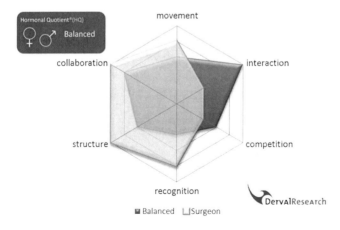

Fig. 3.17 Balanced HQ and Surgeon *Source* Research conducted by DervalResearch on 400 managers and leaders, men and women, from over 20 countries enrolled in executive education programs

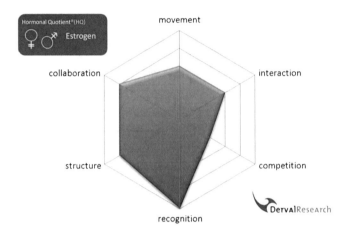

Fig. 3.18 Estrogen HQ and Career Profile *Source* Research conducted by DervalResearch on 400 managers and leaders, men and women, from over 20 countries enrolled in executive education programs

Estrogen HQ: Chief Information Officer versus Politician

Estrogen-driven men and women perform better in a cooperative context. They don't mind following strict rules as long as they see tangible results. Their sense of service and diplomatic skills makes them ideal ambassadors (Mehta and Josephs 2010) (Fig. 3.18).

Robert-Jan Woltering found the perfect fit as General Manager of a luxury hotel (Fig. 3.19) with a high level of recognition, opportunities to collaborate with the team, and demanding hospitality procedures to follow.

Top jobs: R&D manager, lawyer, accountant, advisor to the board, business analyst, business development director, CEO in IT, CIO, clinic or hotel manager, computer engineer, insurance manager, market researcher, VP compliance & risks, client manager, surgeon.

A role as a Chief Information Officer might also please estrogen people. It gives them the right level of collaboration, and structure. Can be challenging and lacking of recognition, but this depends a lot on the industry and type of organization. A bank, insurance, or industry context with clear procedures and quality control might be ideal (Fig. 3.20).

Jobs to avoid: sales manager, politician, banker, pilot, CRM manager

Estrogen men and women wouldn't particularly enjoy a career as a politician: too much interactions, competition, and movement (Fig. 3.21). This might explain why most politicians are on the testosterone side. Barack Obama shows a very testosterone-driven HQ. Even Angela Merkel, German Chancellor, was a bit of a tomboy and is called the Iron Lady because, like Margaret Thatcher, she studied chemistry. A career as a diplomat might be better suited for estrogen people.

Hormonal Quotient® (HQ) and Ideal Job

Fig. 3.19 Meet Robert-Jan Woltering, Sofitel

On the other hand, our estrogen-driven leaders—Ariane, and Robert-Jan—are into attracting and retaining more clients in their market, building strong relationships with all stakeholders, involving their teams and clients and generating

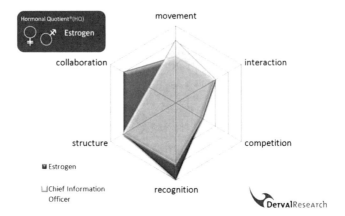

Fig. 3.20 Estrogen HQ and Chief Information Officer *Source* Research conducted by Derval-Research on 400 managers and leaders, men and women, from over 20 countries enrolled in executive education programs

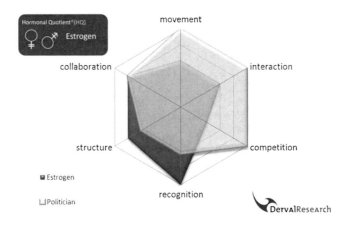

Fig. 3.21 Estrogen HQ and Politician *Source* Research conducted by Derval Research on 400 managers and leaders, men and women, from over 20 countries enrolled in executive education programs

ideas together, and optimizing processes. The extra estrogen seems to encourage co-creation.

We will see in Chap. 4 with the Movie Directors' case that it is possible to succeed in any job no matter which Hormonal Quotient® (HQ) we have, provided there is enough room to make it our own. We will also see how to build winning teams.

Interestingly, our testosterone-driven leaders—Virginie, and Nick—are into creating markets, expanding, being the clear deciders, putting their ideas into motion, working in technological environments. The extra testosterone seems to lead indeed to some "male:male" competition and to trigger the willingness not only to create but to lead people towards uncontestable success.

References

Bennett M, Manning JT, Cook CJ, Kilduff LP (2010) Digit ratio (2D:4D) and performance in elite rugby players. J Sports Sci 28(13):1415–1421 (PubMed PMID:20981610)

Brosnan MJ (2006) Digit ratio and faculty membership: implications for the relationship between prenatal testosterone and academia. Br J Psychol 97(Pt 4):455–466 (PubMed PMID:17018183)

Campbell BC, Dreber A, Apicella CL, Eisenberg DT, Gray PB, Little AC, Garcia JR, Zamore RS, Lum JK (2010) Testosterone exposure, dopaminergic reward, and sensation-seeking in young men. Physiol Behav 99(4):451–456 (PubMed PMID:20026092)

Derval D (2010a) The right sensory mix: targeting consumer product development scientifically. Springer, Berlin

Derval D (2010b) Hormonal Fingerprint and sound perception: a segmentation model to understand and predict individuals' hearing patterns based on OtoAcoustic Emissions, sensitivity to loudness, and prenatal exposure to hormones. In: 30th International Congress of Audiology-ICA 2010. The International Society of Audiology, Sao Paulo

Ebstein RP, Israel S, Lerer E, Uzefovsky F, Shalev I, Gritsenko I, Riebold M, Salomon S, Yirmiya N (2009) Arginine vasopressin and oxytocin modulate human social behavior. Ann N Y Acad Sci 1167:87–102 (PubMed PMID:19580556)

Mehta PH, Josephs RA (2010) Social endocrinology. In: Dunning D (ed) Social motivation. Psychology Press, New York, pp 171–189

Voracek M, Pum U, Dressler SG (2010) Investigating digit ratio (2D:4D) in a highly male-dominated occupation: the case of firefighters. Scand J Psychol 51(2):146–156 (PubMed PMID:19954495)

Chapter 4
Build a Winning Team

> As good as I am, I'm nothing without my band.
> Steven Tyler, Singer of Aerosmith

In this section, we will review the ideal teams, organization, and country depending on our Hormonal Quotient® (HQ). Look into the influence of hormones on entrepreneurship and leadership and analyze the leadership style for each Hormonal Quotient® (HQ) profile. We will share strategies to recruit, motivate, and lead teams. Successful leaders agreed to share their secrets for building winning teams.

Hormones, Entrepreneurship, and Leadership

Are we born entrepreneurs or leaders? What makes us a failure in one country and a successful leader in another one? Here are some answers to these critical questions in our quest for building or being a member of a winning team.

Entrepreneur or Top Manager?

Starting a company or managing one is synonymous with risk taking. Research showed that the influence of prenatal hormones had a direct impact on the level of risk taken by traders and their way of working. Very testosterone ones tended to take more risks, and use more complex algorithms and quantitative analysis—which by the way only leads to better performance in specific cases (Coates and Page 2009).

It is therefore interesting to analyze further the influence of hormones in a sample of 20 women entrepreneurs and 134 male leaders.

The women entrepreneurs were enrolled in an executive education fast-track of 6 months with the objective to turn their business plan into a successful company. 60% of the 20 women were testosterone or very-testosterone driven against for instance 38% only in our previous sample of 67 female shoppers (Fig. 4.1). Among the students, the 6 who successfully created their company were either testosterone or estrogen-driven (Table 4.1).

Fig. 4.1 Meet Magda Carvalho, Patent Attorney

Table 4.1 Women entrepreneurs by Hormonal Quotient® (HQ)

HQ		Enrolled	in %
	Very-testosterone	2	10
	Testosterone	10	50
	Balanced	6	30
	Estrogen	2	10
Total		20	100

Source Research conducted by DervalResearch on 20 women enrolled in ESSEC program for Women Entrepreneurs, in the 2009 spring session

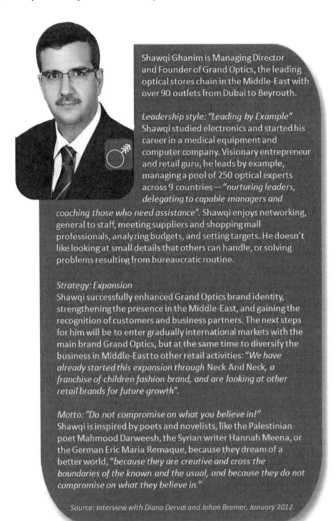

Fig. 4.2 Meet Shawqi Ghanim, Grand Optics

Leaders who are very influenced by estrogen, like Magda Carvalho, Patent Attorney (Fig. 4.1), or by testosterone, like Shawqi Ghanim, Grand Optics (Fig. 4.2), are more likely to become their own boss.

Let's have a closer look now at 134 male leaders and managers enrolled in the Executive MBA program offered by the Robert Kennedy College in Zurich, affiliated with the University of Wales. We considered CEOs and general managers as *top managers*. Interestingly, top managers were over represented among very-testosterone and estrogen-driven students. For instance, estrogen-driven students represented 35.8% of the class and 40.9% of the top managers—1 out of 5 estrogen-driven students is a top manager. Very-testosterone driven students represented 22.4% of

Table 4.2 Top managers by Hormonal Quotient® (HQ)

HQ		Enrolled	in %	Top managers	in %
	Very-testosterone	30	22.4	7	31.8
	Testosterone	31	23.1	4	18.2
	Balanced	25	18.7	2	9.1
	Estrogen	48	35.8	9	40.9
Total		134	100.0	22	100.0

Source Research conducted by DervalResearch on 134 males enrolled in the Robert Kennedy College Executive MBA, in 2010 and 2011

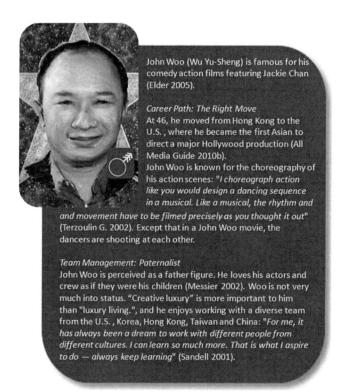

John Woo (Wu Yu-Sheng) is famous for his comedy action films featuring Jackie Chan (Elder 2005).

Career Path: The Right Move
At 46, he moved from Hong Kong to the U.S., where he became the first Asian to direct a major Hollywood production (All Media Guide 2010b).
John Woo is known for the choreography of his action scenes: "*I choreograph action like you would design a dancing sequence in a musical. Like a musical, the rhythm and movement have to be filmed precisely as you thought it out*" (Terzoulin G. 2002). Except that in a John Woo movie, the dancers are shooting at each other.

Team Management: Paternalist
John Woo is perceived as a father figure. He loves his actors and crew as if they were his children (Messier 2002). Woo is not very much into status. "Creative luxury" is more important to him than "luxury living.", and he enjoys working with a diverse team from the U.S., Korea, Hong Kong, Taiwan and China: "*For me, it has always been a dream to work with different people from different cultures. I can learn so much more. That is what I aspire to do — always keep learning*" (Sandell 2001).

Fig. 4.3 Meet John Woo, movie director. *Source* All Media Guide (2010b); Elder (2005); Messier (2002); Sandell (2001); Terzoulin (2002)

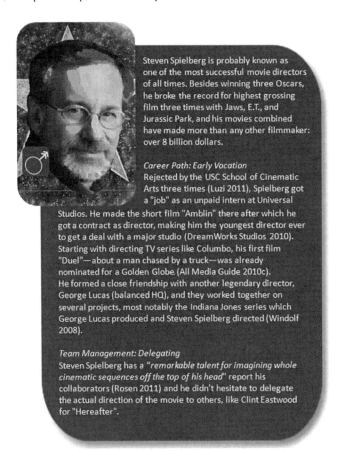

Fig. 4.4 Meet Steven Spielberg, movie director. *Source* All Media Guide (2010c); Dreamworks Studios (2010); Luzi (2011); Rosen (2011); Windolf (2008)

the class and 31.8% of the top managers—1 out of 4 very-testosterone driven students is a top manager. On the other hand, balanced students represented 18.7% of the class but only 9.1% of the top managers (Table 4.2).

In both men and women, the drive for becoming an entrepreneur or a top manager seems greatly linked to our hormonal markup. Let's see if it has also an influence on the leadership style.

Same Job, Different Styles: The Movie Directors' Case

We decided to compare the team management style of four men with each having a different Hormonal Quotient® (HQ) and the same job that gives them enough room to show their own personality: movie director.

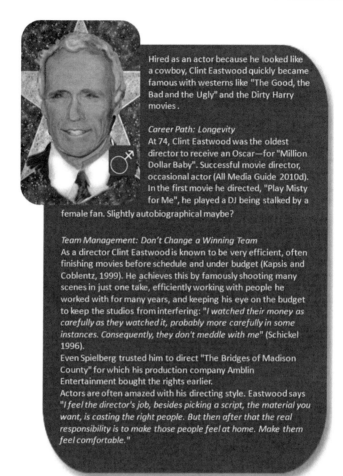

Fig. 4.5 Meet Clint Eastwood, movie director. *Source* All Media Guide (2010d); Kapsis and Coblentz (1999); Schickel (1996)

A movie director manages huge crews—I counted 865 names on the credits of "Indiana Jones and the Kingdom of the Crystal Skull" for instance—(All Media Guide 2010a), a massive budget, a tight deadline, and has on top of it a goal to deliver a movie that makes profits and wins a few awards if possible.

We will see that movie directors like John Woo (Fig. 4.3), Steven Spielberg (Fig. 4.4), Clint Eastwood (Fig. 4.5), and Ron Howard (Fig. 4.7) have their very own style. We will also learn more about the interactions between actors like Meryl Streep (Fig. 4.6) and their movie director.

Among the movie directors, the testosterone-driven John Woo and Steven Spielberg clearly take more risks, changing teams, topics. Ron Howard and Clint Eastwood don't change winning teams. It is interesting though to notice that

Hormones, Entrepreneurship, and Leadership

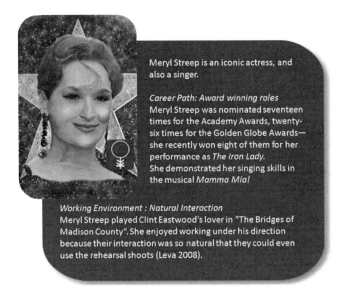

Fig. 4.6 Meryl Streep about Clint Eastwood. *Source* Leva (2008)

Spielberg and Howard delegated the movie direction to Eastwood visibly very effective in that role.

John Woo, became suddenly successful by changing country. Let's see how to find the ideal geographical location or type of organization in which to grow.

Ideal Country and Type of Organization

Understanding differences between countries or companies is key in order to find the right fit for us. Geert Hofstede identified 6 dimensions on which organizations differ: power distance, individualism, masculinity, uncertainty avoidance, long-term orientation, and indulgence (Minkov and Hofstede 2011).

Following mapping between the Hormonal Quotient® (HQ) and the Hofstede model should help pick the right country or type of organization.

In a country or company with a high *power distance* it is tolerated to have some happy fews concentrating the power. Examples cited are Russia, Malaysia, or Romania. A high *individualism* means that individual goals win over collective ones. This would be the case in Australia, the U.S., or the Netherlands. Ideal for people who need *recognition*. High *uncertainty avoidance* would be a high aversion to risk and has been observed in Greece, Malta, or Guatemala. Not to recommend for people into *movement*! *Long term orientation* is more frequent in China, Germany, or the Ukraine. *Indulgence* means that people are looking for fun in life as opposed to following strict rules. Among them, countries like Nigeria, Puerto Rico, or Mexico. Probably too chaotic for people into *structure*. High *masculinity* involves that men

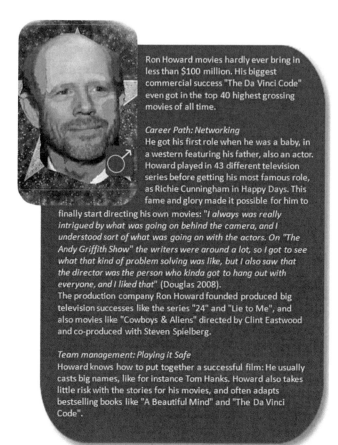

Fig. 4.7 Meet Ron Howard, movie director. *Source* Douglas (2008)

and women are not doing the same tasks. Japan, Hungary, and South Africa are on top. Funnily enough, countries where people are known for being more influenced by testosterone like Finland are low on this masculinity scale (Manning 2002). My interpretation is that in countries where men and women have a similar Hormonal Quotient® (HQ), and consequently similar talents, it is easier for them to succeed in the same type of jobs.

Here a map of the words the most used in the 135 million profiles posted on Linkedin, a leading professional social network, with a focus on the U.S., Canada, the U.K., the Netherlands, Germany, Italy, India, Singapore, and Australia (Fig. 4.8).

Italy scores high on *masculinity* (top 10 out of 111 countries) which explains the importance of a skill like *problem solving* when applying for a job.

It is interesting to see that most professionals from the U.S., Canada, the U.K., the Netherlands, Germany, and Australia, put the word *creative* in their profile. Did these countries attract all the creative people?

Fig. 4.8 "Wordmap" of competencies. *Source* Adapted from Linkedin (2012)

Fig. 4.9 Testosterone HQ, Austria, and the U.S.

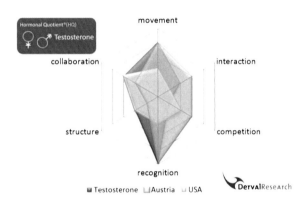

The common point between these 6 countries is that they all score high on *individualism* (in the top 20 out of 111 countries) and low on *power distance* (in the bottom 20 out of 111 countries): a lot of recognition with a minimum inequality.

This is an ideal context for creative people because as we saw in Chap. 2, being creative means having a low binding potential on the dopamine D2 receptors, which means also not being so much into domination and status.

At the other end, India and Singapore score high on *power distance* so that monitorable traits like being *effective* and *track record* are valued. Structured people, with a high binding potential on dopamine D2 receptors, would best fit the bill.

Let's compare how someone with a testosterone-driven Hormonal Quotient® (HQ) would fit in Austria and in the U.S., for instance (Fig. 4.9).

Regarding *power distance*, out of 111 countries analyzed Austria is the one with the lowest score (we put 1) and the U.S. scores 2. In terms of *individualism* the US

Fig. 4.10 Meet Arnold Schwarzenegger, business man, actor, politician. *Source* Lewis (2011)

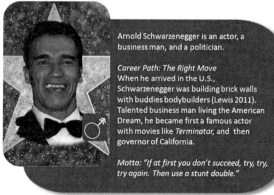

score the highest (we put 5) which is perfect if you are looking for recognition

Fig. 4.11 Meet Bengt Järrehult, "Dr Beng", SCA. *Source* SCA (2011)

while Austria scores 2. *Uncertainty avoidance* is higher in Austria, which means that movement is easier in the U.S. (4 versus 2). *Indulgence* scores and *masculinity* are high in both countries. The mapping for collaboration and interaction is less

Fig. 4.12 Meet Hilary Ellis, talent for change and European PWN

clear. Based on our own observation that chit-chat is less common in Austria, that managers tend to gather input and make decisions on their own, and that there is less competition, we put the three remaining values slightly lower for Austria. U.S. is a very good fit for testosterone people, especially on the recognition, and on the movement side. Lack of movement and recognition can make testosterone people feel constrained in Austria. As *masculinity* is high in the U.S., the fit would be even better for people fueled with testosterone and who actually also are men.

Moving from Austria to the U.S. seems like a good idea for testosterone-driven people. Arnold Schwarzenegger is a very good example (Fig. 4.10).

It is a bit unfortunate though to have to change country to succeed. Let's see if there is a way to be happy in our job at home.

Fig. 4.13 Meet Wai Wong, Sephora

New Strategies to Recruit, Motivate, and Lead Teams

Current recruitment strategies are very formatted. This "clone war" can lead to serious business side effects as we will see in the sales case. We will requalify diversity and share what makes a great leader.

The Clone Wars: The Sales Case

The company was evaluating the opportunity to launch a new mainstream food product. The prototype was first tasted internally, by the sales and marketing team: 300 people worldwide (Scandinavia, Asia, India, Eastern Europe, Western Europe, South America, and the U.S.). Most liked the product and the teams were confident enough to launch it. Unfortunately, it was rejected by the consumers, finding it too bitter (Derval 2010).

Fig. 4.14 « Performance » leaders

	N	Non-taster (%)	Taster (%)
Sales and marketing managers	300	80	20
Market		25	75

Table 4.3 Taste profile comparison (Derval 2009). *Source:* Measurements performed by DervalResearch in 2008 on 300 sales and marketing managers from 20 countries, within the same company

Source Measurements performed by DervalResearch in 2008 on 300 sales and marketing managers from 20 countries, within the same company.

Actually the internal panel was representative in terms of gender, and age, but not in terms of taste. We measured the sensitivity to bitterness of the teams with PTC strips and the surprising results are in the Table 4.3.

Fig. 4.15 « Results » leaders

Among the 300 sales and marketing employees, men and women, from the 25 sales and marketing departments worldwide only 20% perceived the bitter taste, with the lowest number of tasters in the Dutch team (13%) and the highest number in the Scandinavian team (27%), German and Russian having the same proportion of tasters (22%). So 80% of the sales and marketing teams from the firm were non-tasters.

This has to be compared to the market, where non-tasters represent only 25% of the population—the gene is recessive, if you remember from Chap. 1 (Bartoshuk et al. 1994).

In the end, there were 3 times more non-tasters among the sales and marketing teams than among the target population. Measurements we did among other teams such as IT showed a much higher proportion of tasters. This experiment was a real eye-opener because it showed that individuals with similar jobs even have similar tastes, independent of their gender, culture, and country. With the disastrous consequence on business we witnessed. There are more differences between a sales manager and an IT developer, than between a Russian sales manager and a German sales manager.

Fig. 4.16 « Cooperation » leaders

Recruiting the right people is important, as Dr Beng, SCA, confirms (Fig. 4.11), and avoiding hiring clones is strategic.

Recruiting and Motivating the Right People

It is our human tendency to identify with people who look like us and they often have the same Hormonal Quotient® (HQ). The role of HR and leaders is to avoid this clone war—as Hilary Ellis, Talent for Change, confirms (Fig. 4.12)—especially when detecting high potentials and future leaders. The idea is to maintain a hormonal diversity at every level of the company. And when I say hormonal, I do not mean men/women (even though it could be a good start) but testosterone/estrogen. Not

Fig. 4.17 « Process » leaders

always easy, as we repeatedly measured an over-representation of balanced people in large corporations, with people having more extreme HQ being the outcasts.

Trust is the word coming back when talking about team management with Wai Wong, Sephora (Fig. 4.13). Many oxytocin receptors seem required in order to be a good leader.

Let's review the leadership style for each Hormonal Quotient® (HQ) and list the strengths as well as the points to improve.

Hormonal Quotient® and Leadership Style

What Type of Leader are You?

Based on the observation of 500 leaders in 25 countries, we identified 4 types of leaders, based on their Hormonal Quotient® (HQ).

Find out if you are more a « performance » (Fig. 4.14), « results » (Fig. 4.15), « cooperation » (Fig. 4.16), or « process » (Fig. 4.17) oriented type of leader (Derval 2011).

You can check your profile online at www.derval-research.com.

Building a winning team can be achieved in different ways but "hormonal diversity" is definitively a plus for business. Now that we are all very successful in our new job, team, and possibly new country, let's see how to achieve the perfect balance with our personal life.

References

All Media Guide (2010a) Indiana Jones and the Kingdom of the Crystal Skull. *The New York Times*. http://movies.nytimes.com/movie/384336/Indiana-Jones-and-the-Kingdom-of-the-Crystal-Skull/credits. Accessed 2012
All Media Guide (2010b) John Woo—biography. *The New York Times*. http://movies.nytimes.com/person/117248/John-Woo/biography. Accessed 2012
All Media Guide (2010c) Steven Spielberg—biography. *The New York Times*. http://movies.nytimes.com/person/112325/Steven-Spielberg/biography. Accessed 2012
All Media Guide (2010d) Clint Eastwood—biography. *The New York Times*. http://movies.nytimes.com/person/88601/Clint-Eastwood/biography. Accessed 2012
Bartoshuk LM, Duffy VB, Miller IJ (1994) PTC/PROP tasting: anatomy, psychophysics, and sex effects. Physiol Behav 56(6):1165–1171
Coates JM, Page L (2009) A note on trader Sharpe Ratios. PLoS One 4(11):e8036 (PubMed PMID:19946367; PubMed Central PMCID: PMC2776986)
Derval D (2009) Hormonal fingerprint and taste perception. In: 13th annual conference of the society for behavioral neuroendocrinology, East Lansing, p 123
Derval D (2010) The right sensory mix: targeting consumer product development scientifically. Springer, Berlin
Derval D (2011) Réussir son étude de marché en 5 jours. Eyrolles, Paris
Douglas E (2008) Ron Howard revisits Frost/Nixon. ComingSoon.net. http://www.comingsoon.net/news/movienews.php?id=50590. Accessed 2012
DreamWorks Studios (2010) Steven Spielberg. http://www.dreamworksstudios.com/about/executives/steven-spielberg. Accessed 2012
Elder RK (2005) John Woo: interviews (conversations with filmmakers). University Press of Mississippi
Kapsis RE, Coblentz K (1999) Clint Eastwood: interviews. University Press of Mississippi
Leva G (2008) An old fashioned love story: making 'The Bridges of Madison County'. http://www.youtube.com/watch?v=pul3Cp6aTEQ. Accessed 2012
Lewis M (2011) California and Bust. *Vanity Fair*. http://www.vanityfair.com/business/features/2011/11/michael-lewis-201111. Accessed 2012
Linkedin, Buzzwords. http://blog.linkedin.com/2011/12/13/buzzwords-redux. Accessed 2012

Luzi E (2011) The Steven Spielberg three step guide to rejection. http://www.theblackandblue.com/2011/04/05/the-steven-spielberg-three-step-guide-to-rejection. Accessed 2012

Manning JT (2002) Digit ratio: a pointer to fertility, behavior, and health. Rutgers University Press, London

Messier M (2002) The boys in company Woo: Christian Slater, Roger Willie, and John Woo on "Windtalkers". AMC Filmcritic. http://www.filmcritic.com/features/2002/06/the-boys-in-company-woo-christian-slater-roger-willie-and-john-woo-on-windtalkers. Accessed 2012

Minkov M, Hofstede G (2011) The evolution of Hofstede's doctrine. Cross Cult Manag Int J 18(1):10–20

Rosen C (2011) Richard Curtis on 'War Horse,' working with Steven Spielberg and the prescience of 'Love Actually'. AOL Moviefone. http://blog.moviefone.com/2011/12/09/richard-curtis-war-horse-love-actually-interview. Accessed 2012

Sandell J (2001) Interview with John Woo. Bright Lights Film J. http://www.brightlightsfilm.com/31/hk_johnwoo.php. Accessed 2012

SCA (2011) Bengt "Doctor Beng" Järrehult. http://www.sca.com/en/Innovation-at-SCA/this-is-doctor-beng. Accessed 2012

Schickel R (1996) Clint Eastwood: a biography. Knopf, New York

Terzoulin G (2002) John Woo by Gas Terzoulin. *newsfinder.org*. http://www.newsfinder.org/site/readings/john_woo. Accessed 2012

Windolf J (2008) Q&A: Steven Spielberg on Indiana Jones. *Vanity Fair*. http://www.vanityfair.com/culture/features/2008/02/spielberg_qanda200802?currentPage=4. Accessed 2012

Chapter 5
Find the Perfect Balance

> Money doesn't make you happy. I now have $50 million but I was just as happy when I had $48 million.
> Arnold Schwarzenegger, Actor, Businessman, and Politician

In this chapter, we review the ideal lifestyle (Fig. 5.1), hobbies, family life, friends, and partner for each Hormonal Quotient® (HQ) profile. We learn about celebrities' passions (Brad Pitt, Johnny Depp). We also reveal the compatibility between the different Hormonal Quotient® (HQ) profiles.

Hormonal Quotient® (HQ) and Lifestyle

Well being is about finding the lifestyle that is right for us. Appropriate nutrition, and activities are a good start.

Hormones, Thyroid, and Ayurveda

There is a clear link between prenatal hormones and various diseases. An excess of testosterone for instance could lead to osteoarthritis (Zhang et al. 2008; Seeppman and Kapur 2000), schizophrenia (Collinson et al. 2010), prostate cancer (Rahman et al. 2011), myocardial infarction (Kyriakidis et al. 2010), autism spectrum disorder (De Bruin et al. 2009), anorexia nervosa (Coombs et al. 2011), oral cancer (Hopp and Jorge 2011), oral contraceptive side effects (Oinonen 2009), delayed puberty (Manning and Fink 2011), amiothrophic lateral sclerosis (ALS) (Vivekananda et al. 2011), human papillomavirus (Brabin et al. 2008), higher HIV infection rate (Manning et al. 2001). But also eating disorders (Smith et al. 2010), and dependence on alcohol (Kornhuber et al. 2011).

At least now we know what we will die from! I suspect that most scientists and researchers are testosterone driven as there is very little data about men and women who are estrogen driven. From the research and observations conducted by DervalResearch together with the Better Immune System Foundation (www.betterimmunesystem.org), health follows a U-shape curve with very testosterone and

Fig. 5.1 Is it the right lifestyle?

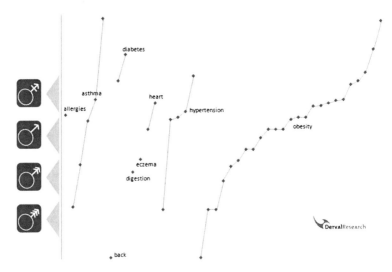

Fig. 5.2 Health issues per Hormonal Quotient® (HQ) in Men. *Source* Research conducted by DervalResearch on 134 males enrolled in the Robert Kennedy college executive MBA, in 2010 and 2011

very estrogen people more likely to have a weaker immune system with associated complications like allergies and auto-immune diseases but also thyroid dysfunctions.

Here are for instance the health issues reported by 134 male managers enrolled in an Executive MBA program (Fig. 5.2). We already had a look at their Hormonal Quotient® (HQ) in Chap. 4 (Table 4.2).

Half of the health issues were reported by estrogen-driven men, who constitute only a third of the students (see Table 4.2). Main disorders include asthma, diabetes,

Table 5.1 Thyroid dysfunctions

	Hypothyroidism	Normal thyroid	Hyperthyroidism
Muscle	Muscle cramps	Normal muscular function	Muscle weakness
Skeleton	Joint pain	Normal maturation	Demineralization
Pulse	Low heart rate	Normal function	High blood rate
Reproduction	Sterility	Normal reproductive ability	Depressed ovarian function
Digestion	Constipation	Normal digestion	Diarrhea
Temperature	Decreased body temperature	Normal temperature	Increased body temperature
Sleep		Normal sleep	Insomnia
Mood	Depression	Normal function	Irritability, restlessness
Appetite	Decreased appetite	Normal appetite	Increased appetite
Skin	Pale, thick	Normal hydration	Flushed, thin
Hair	Thick		Soft

Source Adapted from (Marieb and Koehn 2007, Unit 3 Chapter 16, p. 621)

heart problems, and hypertension (high blood rate)—a sign of hyperthyroidism. Another symptom of a too active thyroid is increased appetite. Obesity—validated by a body mass index higher than 30%—is over-represented among estrogen-driven men.

Balanced men mentioned some heart and weight problems. They are overall fit.

Testosterone-driven men reported digestion problems, eczema, and asthma—signs of a weaker immune system.

Men with a very testosterone Hormonal Quotient® (HQ) were few to report health issues—mainly back problems, asthma, and hypertension—and were also less likely to be overweight. Based on the abundant literature about prenatal testosterone and diseases it sounds like very testosterone-driven men are less likely to be sick, but when they are it sucks.

Both estrogen and very testosterone men reported thyroid problems such as hypertension. Thyroid is a hormone you do not want to mess up with as it affects every cell in the body: stimulating glucose processing, body heat production, and increasing the number of adrenergic receptors in blood vessels (Marieb and Koehn 2007, Unit 3 Chapter 16). Yes, the one putting us in a "fight mode". Dysfunctions can appear later in life with hormonal changes like menopause, for instance. Here are the main symptoms related to the thyroid hormone (TH) (Table 5.1).

Question is, what can we do about it? Probably a little bit on changing the other factors that influence our metabolism: nutrition and activities.

If Ayurveda is considered in Western countries as an alternative medicine, in India—where it was born around 2500 BC—this is a serious medical school, with trained practitioners. In Ayurveda, the idea is that our body is a digestion process with an input and an output. The diagnosis is based on the observation of individuals as a whole, and considers—similarly to what we did in our research on the Hormonal

Quotient® (HQ)—following parameters: immune system, sensitivities, balance, body measurements, diet, digestive capacity, age, and fitness.

In very detailed illustrations, Ayurvedic medicine explains, for instance, how to diagnose diseases based on the appearance of the nails and of the tongue. Cracks on the tongue for instance can suggest a derangement in the colon, and white spots on the nails are clear indicators of zinc or calcium deficiencies. Nutrition advice is also provided, depending on the patient profile. Interestingly, Ayurveda identified 3 types of people: Kapha, who are calm, Pitta, moderated, and Vata, very active (Lad 1985).

The detailed profiles are very strangely similar to the table about thyroid dysfuntions (Table 5.1), with Kapha close to hypothyroidism, Pitta, with a normal thyroid function, and Vata, close to hyperthyroidism—so very testosterone and estrogen men are likely to be Vata. The other interesting parallel is the classification of food. Some foods are *cooling*, like rice, rhubarb, banana, coconut oil, while others are *heating*, like grapes, cheese, radish, beef, having a different effect on digestion and metabolism.

This is worth having a closer look as recent publications confirm the impact of food on hormones receptors. For instance, coffee seems to alter the reaction of estrogen receptors slightly lowering the risk of breast cancer (Li et al. 2011)—time for a coffee break!

In the same way as there is no universally healthy food—if you are allergic to cereals it is not a good idea to eat anything with "whole grains" or "organic wheat" written on it—we all have our own talents to discover.

Discover Your Hidden Talents

Let's see if we can use hormones to unlock some hidden talents. The impact of testosterone on physical and sports abilities has been demonstrated: it quickens the recovery from muscular activities (Pinel 2007), increases targeting ability (Alexander 2006), musical ability (Sluming and Manning 2000), and performance in fencing (Voracek et al. 2006, 2010).

As again little is researched about estrogen people and about women, let's go back to our sample of 67 female shoppers with an overview of their hobbies (Fig. 5.3).

We can see clear patterns by Hormonal Quotient® (HQ). Very testosterone women are into outdoors or violent sports like snowboard, squash, and boxing. Testosterone women a lot into movement sports like swimming, aquarobics (discovering new sports), and aerobics. Balanced women are into nature around trekking, and animals. And estrogen women into indoors focusing on yoga, pets, boardgames and children.

What is even more fascinating is that the nature of hobbies is similar among men. We will learn more about men's hobbies and lifestyle with the examples of Brad Pitt and Johnny Depp.

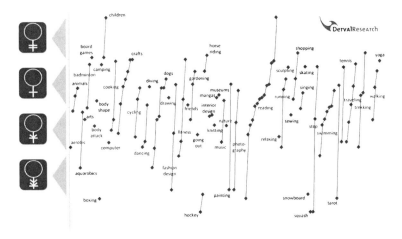

Fig. 5.3 Hobbies per Hormonal Quotient® (HQ) in Women. *Source* Research conducted by DervalResearch on 67 Dutch female shoppers in 2007

Find the Right Lifestyle

Let's have a closer look at the lifestyle of Brad Pitt (Fig. 5.4), who gives us a good example of balance with a rich family life, together with Angelina Jolie and their children, and a successful career.

Knowing our Hormonal Quotient® (HQ) can help us discover some hidden talents and pick the right lifestyle. With an estrogen-driven Hormonal Quotient® (HQ), Brad Pitt's perfect vision of shapes and colors is indeed an asset in the field of design and architecture, as we saw in Chap. 3. It might also explain the choice of colorful chameleons. His love for children and animals is also consistent with the findings on female shoppers. As petting releases oxytocin, our hypothesis is that estrogen people are looking for this "cuddle hormone".

Let's see how hormones influence also our friendships and love relationships.

Hormonal Quotient® (HQ), Friends, and Love

What makes us like or dislike someone? Who would be the perfect buddy or missing half for us? Here are some answers.

Perfect Buddies

Not easy for an estrogen man to make friends at the fitness center if you do not really care about cars and football (this is Johan speaking). Brad Pitt found a good

Fig. 5.4 Meet Brad Pitt, actor. *Source* Schneller (1995); Simon (2011); Singh (2008)

and long lasting friend in Johnny Depp (Fig. 5.5). Let's see what we can learn about this friendship.

Close friend of the Pitt-Jolie couple Johnny Depp is less charmed by their dog Jacques: he reported it jumping him while dogsitting.

With a testosterone-driven Hormonal Quotient® (HQ), Johnny Depp has increased musical ability. As alcohol pleases opioid receptors, Depp, like many testosterone people enjoys some nice wine. Moving to France also highlights some high concentrations of dopamine D4 receptors. His personality complements well Brad Pitt's one and they have definitively more to talk together than cars.

Our Hormonal Quotient® (HQ) can guide our lifestyle, and help explain compatibility between people. We can find the right buddy, but also our missing half.

Hormones and Family Life

Some single people are very happy on their own while others are actively looking for a stable partner. How come?

Unsurprisingly, hormones have again their say in pair bonding and commitment. For instance married men have lower levels of testosterone than unmarried men (Gray et al. 2004), and unmarried men in a relationship have lower levels of testosterone than single men (Mehta and Josephs 2010).

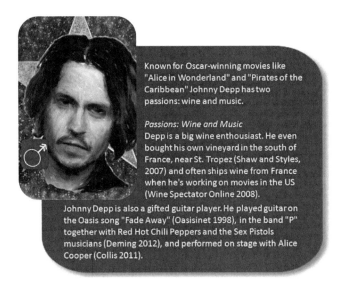

Fig. 5.5 Meet Johnny Depp, actor. *Source* Collis (2011); Deming (2012); Oasisinet (1998); Shaw and Styles (2007); Wine Spectator Online (2008)

Finnish Fertility
Women looking for a fertile guy should travel to Helsinki as Finnish guys have a sperm count higher than any other men (Manning et al. 2000). In general, the more prenatal testosterone in men, the higher their fertility.

Prenatal hormones have a direct influence on the number of our sex partners (Hönekopp et al. 2006), our level of sex arousal (Van den Bergh et al. 2006), and the length of our relationship.

As some might have experienced, being married does not always mean being monogamous. As observed in some birdies, monogamy is linked to the number and location of the oxytocin and vasopression receptors (Ebstein et al. 2009, Ross et al. 2009)—a Siemens Biograph mMR scan should be prescribed before every wedding!

Find Your Missing Half

Here are some tips for finding our missing half (Fig. 5.6) depending on our Hormonal Quotient® (HQ).

Not only the shape of our face can be a precious indicator of our Hormonal Quotient® (HQ), as we saw in Chap. 2, but also the way (Fink et al. 2005) we dance.

Fig. 5.6 Is it the right partner?

A recent experiment video recorded men's dance moves and displayed the clips to a panel of 104 female judges. They were "on average" more attracted to men with a testosterone-driven profile (Fink et al. 2007). What the research does not say is what was the profile of the women judging. Some women favor a male-looking face (Grammer and Thornhill 1994; Scheib et al. 1999) while others prefer a more female-looking male face with a shorter than average lower jaw (Perrett et al. 1998; Penton-Voak et al. 1999).

Once again, what about women from Mars and men from Venus? Luckily, some researchers spent time observing lizards and birdies. In white-throated sparrows, males and females with a white stripe are more aggressive, and males and females with a tan stripe are more care-taking. Opposites apparently attract to form the best teams, capable of both protection and nurturing: 90% of the couples are white-striped males with tan-striped females or white-striped females with tan-striped males. Important to note is that, like in humans, hormones can have an influence on behavior without having necessarily an impact on sexual preferences (Derval 2010).

While researching consumers and behavior of households with DervalResearch, we were amazed to see that people with a given Hormonal Quotient® (HQ) tended to live together with people with a specific Hormonal Quotient® (HQ). We took into account how happy these couples seemed to be (this is of course not easy to evaluate but should not be ignored!). Here are the successful combinations we observed among 50 couples from various nationalities, living together, married or not, with children or without, heterosexual as well as homosexual (Table 5.2).

Based on our observations, here are the winning couples. Estrogen-driven men like Brad Pitt, will be happy with a testosterone or a very testosterone-driven women. Dr Beng, interviewed in Chap. 4, confirms that his third wife is the perfect

Table 5.2 Hormonal Quotient® (HQ) love compatibility grid

Hormonal Quotient®	♂↗	♂↗	♂↗	♂↗
♀↓			♥	♥
♀↓		♥		♥
♀↑	♥		♥	
♀↑	♥	♥		

Source Research conducted by DervalResearch on 50 couples living together.

partner for him. Before he was coupled with women he judges now too similar to himself. Balanced men feel best with balanced women and if they are a bit on the estrogen side they are attracted to very testosterone women, who can take charge and motivate them. Testosterone men such as Johnny Depp are the best partners for testosterone women—they would then be buddies at the same time—or for estrogen women, who they can take care of. Very testosterone men often live with estrogen-driven women or balanced women as they need to have the lead in their fields of interest.

Tracking the length of the face, of the fingers and of any other body part might lead you to the right partner. You can also refer to the jobs and hobbies per Hormonal Quotient® (HQ) to find your missing half.

These compatibilities are also applicable to friendships like between Brad Pitt and Johnny Depp and in the workplace, we can think of the great collaboration between Clint Eastwood and Meryl Streep.

Achieving the perfect balance is all about finding the right job, enjoyable hobbies, good buddies, and love. Hormones can guide us in this quest.

References

Alexander GM (2006) Associations among gender-linked toy preferences, spatial ability, and digit ratio: evidence from eye-tracking analysis. Arch Sex Behav 35(6):699–709 (PubMed PMID:16708283)

Brabin L, Roberts SA, Farzaneh F, Fairbrother E, Kitchener HC (2008) The second to fourth digit ratio (2D:4D) in women with and without human papillomavirus and cervical dysplasia. Am J Hum Biol 20(3):337–341 (PubMed PMID:18203126)

Collis C (2011) Johnny Depp talks about filming his Keith Richards documentary: 'It was intense'. Entertainment Weekly. http://music-mix.ew.com/2011/01/14/johnny-depp-keith-richards-documentary. Accessed 2012

Collinson SL, Lim M, Chaw JH, Verma S, Sim K, Rapisarda A, Chong SA (2010) Increased ratio of 2nd to 4th digit (2D:4D) in schizophrenia. Psychiatry Res 176(1):8–12 (PubMed PMID:20083312)

Coombs E, Brosnan M, Bryant-Waugh R, Skevington SM (2011) An investigation into the relationship between eating disorder psychopathology and autistic symptomatology in a non-clinical sample. Br J Clin Psychol 50(3):326–338 (PubMed PMID:21810110)

De Bruin EI, De Nijs PF, Verheij F, Verhagen DH, Ferdinand RF (2009) Autistic features in girls from a psychiatric sample are strongly associated with a low 2D:4D ratio. Autism. 13(5):511–521 (PubMed PMID:19759064)

Deming M (2012) P Biography. All Music Guide. http://www.allmusic.com/artist/p-p159081. Accessed 2012

Derval D (2010) The Right Sensory Mix: Targeting Consumer Product Development Scientifically. Springer, Berlin

Ebstein RP, Israel S, Lerer E, Uzefovsky F, Shalev I, Gritsenko I, Riebold M, Salomon S, Yirmiya N (2009) Arginine vasopressin and oxytocin modulate human social behavior. Ann N Y Acad Sci 1167:87–102

Fink B, Grammer K, Mitteroecker P, Gunz P, Schaefer K, Bookstein FL, Manning JT (2005) Second to fourth digit ratio and face shape. Proc Biol Sci 272(1576):1995–2001 (PubMed PMID:16191608; PubMed Central PMCID: PMC1559906)

Fink B, Seydel H, Manning JT, Kappeler PM (2007) A preliminary investigation of the associations between digit ratio and women's perception of men's dance. Personality Individ Differ 42(2):381–390

Grammer K, Thornhill R (1994) Human (Homo sapiens) facial attractiveness and sexual selection: the role of symmetry and averageness. J Comp Psychol 108(3):233–242

Gray PB, Flynn Chapman J, Burnham TC, McIntyre MH, Lipson SF, Ellison PT (2004) Human male pair bonding and testosterone. Hum Nat 15:119–131

Hönekopp J, Voracek M, Manning JT (2006) 2nd to 4th digit ratio (2D:4D) and number of sex partners: evidence for effects of prenatal testosterone in men. Psychoneuroendocrinology. 31(1):30–37 (PubMed PMID:16005157)

Hopp RN, Jorge J (2011) Right hand digit ratio (2D:4D) is associated with oral cancer. Am J Hum Biol. 23(3):423–425 (PubMed PMID:21445935)

Kornhuber J, Erhard G, Lenz B, Kraus T, Sperling W, Bayerlein K, Biermann T, Stoessel C (2011) Low digit ratio 2D:4D in alcohol dependent patients. PLoS One 6(4):e19332 (PubMed PMID:21547078; PubMed Central PMCID: PMC3081847)

Kyriakidis I, Papaioannidou P, Pantelidou V, Kalles V, Gemitzis K (2010) Digit ratios and relation to myocardial infarction in Greek men and women. Gend Med 7(6):628–636 (PubMed PMID:21195362)

Lad V (1985) Ayurveda: the science of self healing. Lotus Press, Twin Lakes

Li J, Seibold P, Chang-Claude J, Flesch-Janys D, Liu J, Czene K, Humphreys K, Hall P (2011) Coffee consumption modifies risk of estrogen-receptor negative breast cancer. Breast Cancer Res 13(3):R49 (PubMed PMID: 21569535)

Manning JT, Barley L, Walton J, Lewis-Jones DI, Trivers RL, Singh D, Thornhill R, Rohde P, Bereczkei T, Henzi P, Soler M, Szwed A (2000) The 2nd:4th digit ratio, sexual dimorphism, population differences, and reproductive success: evidence for sexually antagonistic genes? Evolut Hum Behav 21:163–183

Manning JT, Henzi P, Bundred PE (2001) The ratio of 2nd to 4th digit length: a proxy for testosterone, and susceptibility to HIV and AIDS? Med Hypotheses 57(6):761–763 (PubMed PMID:11918443)

Manning JT, Fink B (2011) Is low digit ratio linked with late menarche? Evidence from the BBC internet study. Am J Hum Biol 23(4):527–533 (PubMed PMID:21547980)

Marieb EN, Hoehn KN (2007) Human anatomy and physiology. Pearson, Benjamin Cummings
Mehta PH, Josephs RA (2010) Social endocrinology. In: Dunning D (ed) Social motivation. Psychology Press, New York, pp 171–189
Oasisinet (1998) "Don't Go Away" Japan Single Release. http://www.oasisinet.com/NewsArticle.aspx?n=176. Accessed 2012
Oinonen KA (2009) Putting a finger on potential predictors of oral contraceptive side effects: 2D:4D and middle-phalangeal hair. Psychoneuroendocrinology 34(5):713–726 (PubMed PMID:19131172)
Penton-Voak IS, Perrett DI, Castles DL, Kobayashi T, Burt DM, Murray LK, Minamisawa R (1999) Menstrual cycle alters face preference. Nature 399(6738):741–742
Perrett DI, Lee KJ, Penton-Voak I, Rowland D, Yoshikawa S, Burt DM, Henzi SP, Castles DL, Akamatsu S (1998) Effects of sexual dimorphism on facial attractiveness. Nature 394(6696):884–887
Pinel JPJ (2007) Basics of biopsychology. Allyn & Bacon, Boston
Rahman AA, Lophatananon A, Stewart-Brown S, Harriss D, Anderson J, Parker T, Easton D, Kote-Jarai Z, Pocock R, Dearnaley D, Guy M, O'Brien L, Wilkinson RA, Hall AL, Sawyer E, Page E, Liu JF (2011) UK Genetic Prostate Cancer Study Collaborators, British Association of Urological Surgeons' Section of Oncology, Eeles RA, Muir K. Hand pattern indicates prostate cancer risk. Br J Cancer 104 (1):175–177 (PubMed PMID:21119657; PubMed Central PMCID: PMC3039824)
Ross HE, Freeman SM, Spiegel LL, Ren X, Terwilliger EF, Young LJ (2009) Variation in Oxytocin Receptor Density in the Nucleus Accumbens Has Differential Effects on Affiliative Behaviors in Monogamous and Polygamous Voles. J Neurosci 29(5):1312–1318
Scheib JE, Gangestad SW, Thornhill R (1999) Facial attractiveness, symmetry and cues of good genes. Proc Biol Sci 266(1431):1913–1917
Schneller J (1995) Brad Attitude. Vanity Fair. http://www.vanityfair.com/hollywood/features/1995/02/brad-pitt-199502. Accessed 2012
Seeppman P, Kapur S (2000) Schizophrenia: More dopamine, more D_2 receptors. Proc Natl Acad Sci U S A 97(14):7673–7675
Shaw L, Styles O (2007) Johnny Depp buys girlfriend vineyard estate in France. *Decanter Magazine*. http://www.decanter.com/news/wine-news/486159/johnny-depp-buys-girlfriend-vineyard-estate-in-france. Accessed 2012
Simon B (2011) Angelina Jolie: Behind the camera. CBS 60 Minutes. http://www.cbsnews.com/video/watch/?id=7389746n&tag=contentMain;cbsCarousel. Accessed 2012
Singh A (2008) Brad Pitt turns architect and designs Dubai hotel. *The Telegraph*. http://www.telegraph.co.uk/news/celebritynews/2066592/Brad-Pitt-turns-architect-and-designs-Dubai-hotel.html. Accessed: 2011
Sluming VA, Manning JT (2000) Second to fourth digit ratio in elite musicians: Evidence for musical ability as an honest signal of male fitness. Evolut Hum Behav 21(1):1–9
Smith AR, Hawkeswood SE, Joiner TE (2010) The measure of a man: associations between digit ratio and disordered eating in males. Int J Eat Disord 43(6):543–548 (PubMed PMID:19718667)
Van den Bergh B, Dewitte S (2006) Digit ratio (2D:4D) moderates the impact of sexual cues on men's decisions in ultimatum games. Proc Biol Sci 273(1597):2091–2095. (PubMed PMID:16846918; PubMed Central PMCID: PMC1635480)
Vivekananda U, Manjalay ZR, Ganesalingam J, Simms J, Shaw CE, Leigh PN, Turner MR, Al-Chalabi A (2011) Low index-to-ring finger length ratio in sporadic ALS supports prenatally defined motor neuronal vulnerability. J Neurol Neurosurg Psychiatry 82(6):635–637 (PubMed PMID:21551173)
Voracek M, Reimer B, Ertl C, Dressler SG (2006) Digit ratio (2D:4D), lateral preferences, and performance in fencing. Percept Mot Skills 103(2):427–446 (PubMed PMID:17165406)
Voracek M, Reimer B, Dressler SG (2010) Digit ratio (2D:4D) predicts sporting success among female fencers independent from physical, experience, and personality factors. Scand J Med Sci Sports 20(6):853–860 (PubMed PMID:19843265)

Wine Spectator Online (2008) Unfiltered: On Set, Johnny Depp Longs for Wines From His Adopted Home. http://www.winespectator.com/webfeature/show/id/Unfiltered-On-Set-Johnny-Depp-Longs-for-Wines-From-His-Adopted-Home_4109. Accessed 2012

Zhang W, Robertson J, Doherty S, Liu JJ, Maciewicz RA, Muir KR, Doherty M (2008) Index to ring finger length ratio and the risk of osteoarthritis. Arthritis Rheum 58(1):137–144 (PubMed PMID:18163515)

Conclusion

Diana: It took me 35 years to find out that I was made for research, that I should be my own boss, that I could manage people but only if they were talented, that Amsterdam was the best place for me to live, that vegetables and cereals were not my friends, that I was a gifted bass player, that karate was the best sport for me provided I had a personal instructor, and that I was allergic to cats. And that all that was due to the influence of prenatal hormones.

And most importantly I found love.

Johan: It took me 30 years to realize that I like 3D design and architecture, that team sports are not for me (or the contrary), that I prefer to collaborate rather than to impose my view, that watching funny cat pictures on the Internet is not a waste of time, that I should go on listening to metal music but not try to play it myself, and that I really enjoy structure.

And I found love too!

We hope that this book will save you a bit of time and effort!

You are welcome to share your experience and thoughts on the website www.derval-research.com. You will find there more tests, measuring discs, workshops, and resources about the Hormonal Quotient® (HQ).

Warm greetings,
Diana & Johan

Printed by Publishers' Graphics LLC
MO20120614